T0231685

PERSPECTIVES ON STRATEGIC ENVIRONMENTAL ASSESSMENT

PERSPECTIVES ON STRATEGIC ENVIRONMENTAL ASSESSMENT

Edited by

MARIA ROSÁRIO PARTIDÁRIO
RAY CLARK

LEWIS PUBLISHERS
Boca Raton London New York Washington, D.C.

Library of Congress Cataloging-in-Publication Data

Catalog record is available from the Library of Congress

Visit the CRC Press Web site at www.crcpress.com

© 2000 by CRC Press LLC
Lewis Publishers is an imprint of CRC Press LLC

No claim to original U.S. Government works
International Standard Book Number 1-56670-360-3
Library of Congress Card Number 99-044581
3 4 5 6 7 8 9 0
Printed on acid-free paper

Preface

On April 16, 1969, at a public hearing on environmental policy before the United States Senate Committee on Interior and Insular Affairs, I urged that proposed legislation have "an action-forcing operational aspect — that the Congress should require the Federal agencies, in submitting proposals, to contain within the proposals an evaluation of their effect upon the state of the environment." This recommendation was incorporated in Section 102(2) (c) of the National Environmental Policy Act of 1969 (NEPA), requiring that "all agencies of the Federal Government shall include in every recommendation or report on proposals for legislation and other major Federal actions significantly affecting the quality of the human environment, a detailed statement by the responsible official on

   (i) the environmental impact of the proposed action
  (ii) any adverse environmental effects which cannot be avoided should the proposal be implemented,
 (iii) alternatives to the proposed action,
 (iv) the relationship between local short-term uses of man's environment and the maintenance and enhancement of long-term productivity, and
  (v) any irreversible and irretrievable commitments of resources which would be involved in the proposed action should it be implemented."

This provision, which I described as "action-forcing," was clearly intended to require an assessment of the environmental effects of proposed government action for the purpose of implementation of the principles enumerated in the congressional Declaration of National Environmental Policy Section 101(b) of the statutory Act stating that — "it is the continuing responsibility of the Federal Government to use all practicable means consistent with other essential considerations of national policy, to improve and coordinate Federal plans, functions, programs, and resources the end that the Nation may

(1) fulfill the responsibilities of each generation as trustee of the environment for succeeding generations;
(2) assure for all Americans safe, healthful, productive, and aesthetically and culturally pleasing surroundings;
(3) attain the widest range of beneficial uses of the environment without degradation, risk to health or safety, or other undesirable and unintended consequences;
(4) preserve important historic, cultural, and natural aspects of our national heritage, and maintain, wherever possible, an environment which supports diversity, and variety of individual choice;

(6)   enhance the quality of renewable resources and approach the maximum attainable recycling of depletable resources."

The action-forcing mechanism was never intended to generate long tomes or delay projects or cost millions of dollars. It was intended to help decisionmakers choose among different strategic directions. Contrary to the case law that has developed over the years, NEPA's purpose was never the writing of impact statements, but to provide an analysis and an inducement to ecological rationality. To use EIA as it was intended requires leadership commited to the objectives and subjecting budgets, strategies, programs and plans to that end.

The concept of informing the consequences of actions impacting the physical environment previously had been expressed in check-lists and cautionary guides. The American legislation, however, appears to have been the first to mandate by law an analysis and assessment of environmental impacts. Need for such an analysis and evaluation was widely recognized, as evidenced by the rapid and widespread adoption of environmental assessment by governments around the world. Establishment of the International Association for Impact assessment in 1981 represents an extraordinarily rapid professionalization of impact assessment of world-wide dimensions.

Environmental assessment was, and is, an innovative aspect of public policy formation. But innovations in government are seldom easily incorporated into existing structures and procedures. This is especially so when a new administrative requirement necessitates development of appropriate criteria, scope, methods, and procedures. Satisfaction of these needs requires time and experience; and their attainment often is confronted by bureaucratic and political incomprehension and resentment.'Importantly it requires a broadened perspective, including additional scientific, technical, and quality of life considerations. Environmental protection and enhancement often obstructs influential economic and political plans and interests and may require change in the expectations and behavior of ordinary people.

A consequence of this reality may be, a mismatch between the assessment process, the preferences of political decisionmakers, and the implementation of declared policy environmental. A way around such conflict has been to regard impact assessment as an informational technique, the findings of which need no more than review by the relevant agency. The Supreme Court of the United States has regarded the environment assessment requirement as no more than information which a Federal agency is required to consider, but which need not affect its policy decision.

This separation of assessment and decision is contrary to the intentions of the framers of NEPA. The action to be forced by the impact analysis, statement, or assessment was implementation of the principles set forth in the Declaration. The connection is clearly evident in the legislative history of the act and logically inferential from the text of the Act itself. Unfortunately, the language of the Act does not establish an incontrovertible link between its two principal sections (101 and 102). Nevertheless, NEPA has had a significant effect upon planning and decision-making in the U.S. Federal agencies. Similar statutes have been adopted in more than half of the States.

The impact assessment principal was endorsed by the United Nations Conference on Environment and Development (1992). But the pressures of population and economic growth have compromised environmental conservation efforts. Strategic planning and implementation are needed to enable the impact assessment process to realize its purpose. More than improved technique is necessary. Somehow the connection between substantive principles and analytic procedures must be established of the goal of a sustainable self-renewing environment, broadly conceived, is to be attained. The assessment process must be clearly linked to the processes of planning and decision-making.

This book constitutes a report on how this goal is being approached in a number of countries and sub-national jurisdictions. It is a valuable and timely contribution. Many of its authors are prominent thinkers and practitioners in the field of EIA. Dr. Partidário and Mr. Clark, the co-editors, are well-respected around the world for their other contributions to the development of impact analysis. I am personally hopeful that new ways of approaching EIA will evolve as move into a new century. There is ample reason to believe that an emerging number of changes in the Earth's environment will become threatening in the coming century. Present day society risks doing too little too late to prevent changes in the Earth's atmosphere, climate, water, soils, and biota that could endanger the sustainability of life on Earth. Notable achievements in science and technology may be insufficient to prevent the degradation of the planet if the bio-physical basis of life continues to be diminished and impaired. Strategies for impact assessment should imply and lead to strategies for policy in action. The World Scientist's Warning to Humanity (Union of Concerned Scientists, 1992) details the environmental problems facing the world. Through strategic impact assessment humanity will hopefully obtain the means and the will to effectively address them.

Lynton K. Caldwell
Professor of Public and
Environmental Affairs, Indiana University

Note: Professor Caldwell was a major participant in drafting the United States National Environmental Policies Act of 1969. He is author of a recent book entitled *The National Environmental Policy Act: Agenda for the Future* (1998) and is an Honorary Member of the International Association for Impact Assessment.

About the Contributors

Pierre André is assistant professor of environmental studies at the Department of Geography, University of Montreal, and is teaching integrated EIA and developing research on environmental perception and public participation in EA.

Ronald Bass is a principal with Jones & Stokes Associates, an environmental consulting firm with offices throughout the western U.S. As an environmental planner and attorney, he has extensive experience managing EIAs.

Hugh Benevides is a Canadian lawyer with a special interest in environmental law and policy. From 1997–1999, he was legislative assistant to Charles Caccia, Canadian Member of Parliament, and is currently based in New Delhi, India.

Clare Brooke works for the Royal Society for the Protection of Birds and the U.K. BirdLife partner, working on environmental assessment (EA) and strategic environmental assessment (SEA), implementing and developing methodologies for SEA, particularly, at the regional level.

A. Lex Brown is professor and head of the School of Environmental Planning at Griffith University, Brisbane, Australia, and has experience in environmental planning and management tools to interface the environmental scientist and the planning professions.

Ray Clark is the Principal Deputy Assistant Secretary of the Army for Installations and Environment. Previously, he was Associate Director for the Council on Environmental Quality in the Executive Office of the President of the United States. Ray has authored numerous papers and co-edited a book on environmental policy.

Jean-Pierre Gagné is professor and director of the School of Audiology and Speech-Language Pathology. He is involved in different projects related with noise distribution, perception, and nuisance.

Anna Hamilton is a North American ecologist with specific expertise in aquatic biology, water quality, estuarine and wetland management, and large-scale monitoring programs, and she is involved in research on applied environmental problems and impact assessments.

Stephen Hazell is a principal and general counsel of Marbek Resource Consultants, where he advises clients in Canada, the Americas, and Southern Africa on environmental law, policy, and programs.

Albert Herson is president of Jones & Stokes Associates, an environmental consulting firm with offices throughout the western U.S. Mr. Herson is an attorney and planner with considerable experience in environmental and land-use law.

Jennifer Howell is a policy analyst with the Canadian Federal Government. As a geneticist, she has been involved in the environmental, social, and ethical implications of diverse projects, including federal proposals.

Milan Danny Machač is Senior Manager of the Ministry of Environment of the Czech Republic, and is currently involved in environmental conflict management, environmental policy, public participation, and environmental training.

Richard K. Morgan is a senior lecturer on human impacts on the environment and EIA in the Department of Geography, University of Otago, Dunedin, New Zealand, where he recently set up the Centre for Impact Assessment Research and Training.

Filipe V. Moura is an environmental engineer, currently an MSc student at the Technical University in Lisbon, Portugal, focusing on the relationships between environmental impact assessment (EIA) and the transportation sector.

Komeri R. Onorio is a PhD research student at the University of Otago, Dunedin, New Zealand. He was formerly the EIA Officer with the South Pacific Regional Environment Programme (SPREP), Apia, Samoa.

Maria Rosário Partidário is an environmental engineer, professor at the New University of Lisbon, Portugal, developing research and consultancy on environmental planning and management, SEA and EIA, and linking these tools with policy and planning.

Vladimir Rimmel is director of the Regional EIA Centre in the Czech Republic, and is an authorized EIA expert since 1993 and manager of Cleaner Production Association. He graduated with a degree in economy, and is currently involved in EIA practice.

William E. Schramm worked for the Environmental Sciences Division of Oak Ridge National Laboratory and is currently a consultant in Oak Ridge, Tennessee. His research interests include the environment implications of trade liberalization.

Jaye Shuttleworth is director of Environmental Services for the Canadian Department of Foreign Affairs and International Trade, and is extensively experienced in the EA field, previously with the Canadian Environmental Assessment Agency.

Riki Therivel is a research scientist with the University of Oxford's Environmental Change Unit and a visiting professor at Oxford Brookes University's School of Planning, specializing in SEA, EIA, and energy efficiency.

Suzanne Therrien-Richards is environmental science and assessment coordinator with the Department of Canadian Heritage, coordinating the application of EA legislation and policy in Saskatchewan, Manitoba, and the Northwest Territories.

Wil A. H. Thissen is a professor of policy analysis at Delft University of Technology, School of Systems Engineering, Policy Analysis and Management, and does research and teaches, concentrating on the methodology of supporting policy and decision-making processes.

Heli Törttö is a civil engineer and researcher at the Water Resources and Environmental Engineering Laboratory at the University of Oulu, Finland. Her area of expertise is EIA at both project level and strategic level.

Rob Verheem is adjunct-secretary of the EIA Commission of The Netherlands and one of the co-authors of the SEA section of the International Study of the Effectiveness of Environmental Assessment.

Lee Wilson is president of Lee Wilson & Associates (LWA), a water resource and environmental consulting firm headquartered in Santa Fe, New Mexico. He is a certified professional hydrologist, who wrote his first environmental impact statement in 1971.

Keith Wiseman is an environmental planner, and currently manager of Policy, Research, and Review in the Environmental Management Department, Cape Town Metropolitan Council, South Africa.

Lukáš Ženatý is Deputy Director of the Regional Department of the Czech Ministry of the Environment in Ostrava, and is mostly involved in management of environmental problems and public participation in the decision-making process.

Table of Contents

Section VI Forum

Section I

Overview

1 Introduction

Maria Rosário Partidário and Ray Clark

CONTENTS

Over the years, Strategic Environmental Assessment (SEA) has become recognized as a form of environmental assessment that can assist managers and leaders in policy, planning, and programmatic decisions. Decision-makers increasingly believe that SEA has the capacity to influence the environmental and sustainability nature of such strategic decisions, and provide for sound, integrated, and sustainable policy and planning frameworks.

Humans try diligently to foresee the consequences of their actions before they take them. When negative effects can be anticipated, they forestall or modify the action, or willingly accept the consequences.

In 1969, the passage of the National Environmental Policy Act in the United States catalyzed the development of analytical tools to predict the environmental and social implications of decisions. Since then, more than 80 countries have passed legislation that requires a full accounting of the likely impacts of decisions.

The International Association for Impact Assessment (IAIA) was established in this context. With more than 2500 members in over 95 countries

IAIA's mission is to advance the innovation, development and dissemination of best practice in environmental impact assessment, management and policy throughout the world. IAIA supports individuals and organizations involved in these and related disciplines by providing a forum for the exchange of ideas and opportunities for collaboration. Its primary purpose is development of international and local capacity to make wise decisions regarding the anticipation, planning and management of environmental change — in terms of ecological and human consequences — in order to enhance the quality of life for all. IAIA promotes ecologically sustainable justice and equitable development and is committed to environmental justice and the preservation of human rights.

Annually, IAIA host conferences to explore new ideas. In 1997, the United States hosted the 17th Annual Conference of IAIA in New Orleans, Louisiana. One of the

1-56670-360-3/00/$0.00+$.50
© 2000 by CRC Press LLC

major themes of this conference was to explore this emerging idea that SEA is a better way of looking at development proposals.

Considerable work has developed in SEA, representing diverse perspectives and approaches. Investigation on common frameworks and key elements of good practice led to the useful compilation of case-experiences, in the form of examples of applications of SEA and also as procedural approaches.[1-3]

Defining SEA is not easy. Few have attempted to venture further than to say that SEA is the environmental assessment of policies, plans, and program.[1,4] Most say that SEA is merely Environmental Impact Assessment (EIA) at any level above the project level.

The complexity associated with the idea, and the need to stress the continuous, proactive, and integrated nature of SEA motivated yet another formulation:[5]

> SEA is a systematic, on-going process for evaluating, at the earliest appropriate stage of publicly accountable decision-making, the environmental quality, and consequences, of alternative visions and development intentions incorporated in policy, planning, or program initiatives, ensuring full integration of relevant biophysical, economic, social and political considerations.

Despite academicians' general acceptance of the idea , the formal public adoption of SEA has been experiencing some difficulties. The complexity of the processes associated to SEA, the consequent need for additional resources, the fact that it is often indicated as having little added value in relation to project's EIA, are some of the factors that are limiting a more extensive adoption of SEA.

Practitioners of SEA frequently find themselves searching for the indisputable arguments that can justify the added value of SEA, particularly where SEA is not yet supported by legal requirements, or where the scope of application is still rather undefined and conflicting with other evaluation procedures, such as traditional EIA.

To contribute to a clarification on the evident, and potential, benefits of SEA adoption, a specific workshop was organized at the occasion of the 17th Annual Conference of IAIA in New Orleans, Louisiana. Many impact assessment professionals sat together for two days to discuss the issue. A number of papers were prepared for the purpose, under the general topic: "What would you sell about SEA that would convince the skeptic about the benefits (environmental and economic) of SEA?"

This book contains a selection of papers presented at this conference, particularly to the SEA conference stream. Papers were subsequently reviewed, and updated, by the respective authors. The title of the book *Perspectives on Strategic Environmental Assessment* was chosen to reflect a set of contributions whose main shared elements are:

- an IAIA conference focus
- the workshop topic: benefits of SEA and how to convince the skeptic

As such, rather than establishing any one form of designing SEA, the book intends to reflect the wide range of perspectives that currently assist the development and evolution of SEA.

UNDERSTANDING SEA

As the world's attention turns to sustainability and the considerations of cumulative effects, the concept of SEA has taken on more significance and urgency. Whether through highly structured and rationalized processes, deeply regulated, or as a process of changing minds and forms of decisions, through better planning and policy-making practices, SEA is seen as providing an adequate context and rationale for sound and integrated decision-making, within which to address synergistic and long-term induced effects (Figure 1.1). Importantly, the practitioners of SEA have recognized that EIA must understand and integrate development principles and not focus entirely on environmental issues.

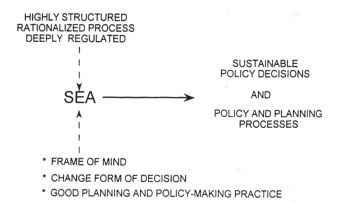

FIGURE 1.1 Approaches to SEA: a name in a process of change, towards sustainability.

Consistent with the nature of an SEA system is, therefore, its potential capacity to contribute to the achievement of sustainability aims. SEA can play a role toward sustainability if the following conditions are met:

- a policy framework is in place, establishing the articulation across sectoral policies and institutional contexts
- credible and feasible alternatives that allow evaluation of a decision based on comparable rather than in absolute values
- recognition that uncertainty characterizes any policy and planning decision
- simple though pragmatic indicators that can assist monitoring of the decisions to determine the actual effects
- good communications mechanisms to ensure that all partners in the SEA process are adequately involved and their perspectives contemplated

In addition, policy-makers must empower the analyst to study alternatives that may be broader than the policy-maker's authority, i.e., to begin coordination with the public and other agencies before there is a firm proposal. Whether or not these conditions are in place is also a key element in the characterization of SEA systems.

Current evidence with SEA shows, however, that there can be many different ways to, and forms of, achieving SEA generally acknowledged objectives — almost as many as the various forms and contexts of decision-making.

SEA is generally presented as the assessment tool that addresses the environmental implications of decisions made above project level. This simplistic way of suggesting the concept of SEA is generating significant controversy, given the enormous range of decision scales and development implications involved in different jurisdictions (Figure 1.2).

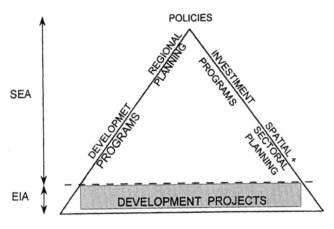

FIGURE 1.2   Can SEA be one, and yet be valid from policies to programs?

The reasons have been more extensively argued elsewhere[5] but are essentially twofold: one is the meaning of the word strategic, which means different things in different places; the other is the enormous range of decision scales and development implications, from policies to programs, that are all encapsulated in the strategic concept.

Sometimes SEA is exclusively associated to the development of high level policy by elected and appointed officials at a government level.[6] In other occasions, the concept is extended to refer to the establishment of policy frameworks for subsequent development consents (U.K. development plans). However, it may also refer to the process of evaluating groups of actions related geographically or having similarities of project type, timing, media or technological character (U.S. programmatic EIS).

Considering the variety of potential applications, as stated in the previous paragraph, with imagination and flexibility one can design a tool that can be adapted, and effectively responsive, to such a wide range of rationalisms, decision levels, and associated decision-making systems (Figure 1.3).

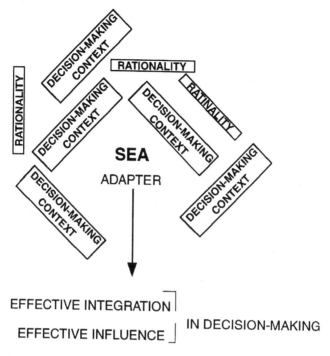

FIGURE 1.3 There is not just one form of SEA. SEA must be adapted to all forms of decision-making and decision rationalities.

FORMS OF SEA

At the core of SEA methodologies are the familiar tools and techniques of EIA. The new form called SEA is about timing, scope and breadth of policy assessment, planning and project EIA. The rate to which SEA is patterned after policy and plan evaluation, or project EIA, depends on the importance and strengths of each respective procedural model, with respect to national and regional decision-making. As a consequence, approaches to SEA commonly exhibit elements that can be identified with top-down (after policy or plan evaluation) or bottom-up (after project EIA) approaches.[5]

Under the circumstances, SEA must be absolutely tailor-made to the kind of decision at stake, and the nature of the decision-making process in place.[7] If this is true for project EIA, it is even more true for SEA, as policy and planning decisions tend to be greatly intuitive, with less detailed information, as well as more incremental than decisions taken at project level.

It is increasingly argued that planning and policy evaluation approaches exist that satisfy main elements of SEA, without necessarily being labeled as SEA. Evidence on this fact is still quite dispersed and difficult to verify. An example is offered by Ray Clark in Chapter 2.

In fact, in many parts of the world, existing procedures require environmental considerations in policy, planning, or program development. Some of these procedures require commitments, site suitability studies and appraisal of optional locations, public consultation, and a number of aspects that are similar to the SEA process. In most situations, no SEA label is attached to those processes.

Given the varied planning systems that exist in the world, any attempts to rationalize SEA into highly contained perspectives, that only fit into some decision-making frameworks, is not helpful. Such dogmatic approaches limit the potential to influence decision-making. This book will argue that it is the SEA application of principles, and the good practice reflected in environmentally sound decisions, that really matter.

However, there is one important aspect about SEA that must not be forgotten when generalizing the SEA nature of some existing appraisals: the fact that SEA, being an environmental assessment process, can only be effective if a consistent and systematic approach is in place.

Basic elements for effective SEA, and success factors in SEA (based on Ref. 8), can thus be suggested at this point (Tables 1.1 and 1.2).

TABLE 1.1
Basic Elements in Effective SEA

Clear requirements (legal basis, administrative order, or policy)
Public participation requirements and public reporting
Well-established process, including main steps
Guidelines for good practice
Assistance or support in each case (private or governmental consultation)
Independent oversight and review of the implementation and performance

---

TABLE 1.2
Success Factors in SEA

Clear environmental policy objectives
Good state of the environment reporting
Well-structured planning process, and proponent commitment and accountability
Multiple organizations that work together
Objectives, criteria, and quality standards framework
   - to assess proposal need and justification
   - to assess environmental effects (losses/changes)
Resources availability
Access to information
Public interest and non-governmental organizations' involvement

---

On the other hand, many environmental assessment approaches currently identified as SEAs could be questioned as to their actual strategic nature. Often, it is not easy to decide on the SEA or project's EIA nature of certain environmental approaches, or even if we are dealing only with better environmental planning practices. It can be argued that there is no need for the establishment of new procedures such as SEA where situations could be dealt with by other forms of well-acknowledged environmental assessment tools, such as project EIA.

In this context, and as represented in Figure 1.4, it can be questioned whether the scope of SEA should be limited to just one level of decision-making or, instead, if it can actually be conceptualized to respond effectively to such a wide scope of decision levels, from policy to programmatic levels. There is no doubt SEA can be quite instrumental in the tiering focus of environmental assessment, however, there may be difficulties associated with the establishment of SEA that are related to such enormous demand pending on SEA performance capacities.[9]

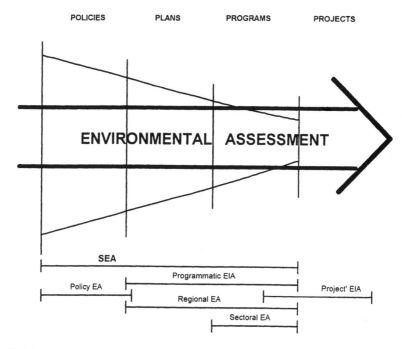

FIGURE 1.4   Focusing environmental assessment across sequential decision-making levels.

Perhaps the content of this book, and the various examples presented, may contribute to the re-thinking of SEA and consider whether the existing differences between policy, plan, and program should in fact be acknowledged by the environmental assessment approach to be adopted. If so, this justifies distinguishing between forms currently known as SEA approaches.

FRAMEWORK AND OVERVIEW OF THE BOOK'S CHAPTERS

As indicated, this book is based on papers presented to the IAIA'97 Strategic Environmental Assessment workshop. Papers presented at that session were reviewed and updated by the respective authors.

This book is therefore a collection of different contributions by professionals from various geographic latitudes who are making use of SEA as an environmental assessment tool. By bringing together various different experiences concerning the application of SEA, the editor and contributors believe that this book will provide a range of useful lessons for future application and improvement of SEA as a policy, and planning decision-making support tool.

For the purpose of greater consistency across the various chapters, each author was asked to organize the chapter in a way that would particularly highlight the benefits, difficulties, lessons learned, and success stories with each SEA case. Although not using a rigid framework, chapters roughly address the following issues:

1. What are the big issues about SEA?
2. Why promote SEA?
3. What works well, and what does not, in each case?
4. Ways forward for improvement in similar contexts

Apart from the first and final sections, respectively, the introduction and conclusions chapters, the remaining 17 chapters of this book are organized into 4 main sections:

1. Section II: SEA in a Sustainability Perspective
2. Section III: Procedural Approaches to SEA at a Policy Level
3. Section IV: Procedural Approaches to SEA at a Plan and Program Level
4. Section V: Methods and Applications — The Role of SEA in Decision-Making

Chapters 2 and 3, in Section II, stress the role of SEA in bridging the sustainability perspective in two different forms: a more political (Chapter 2) and a more analytical (Chapter 3) form.

Chapters 4 to 8, in Section III, develop SEA in a policy context, either by arguing its legal development (Chapter 4), its procedural approaches (Chapters 5 and 6), its actual role at policy level (Chapter 7), or its evaluation requisites in comparison to other policy tools and approaches (Chapter 8).

Section IV offers three chapters that clearly focus the plan and programmatic levels of SEA approach, presenting examples on procedural approaches (Chapters 9, 10, and 11).

Finally, Section V is about case examples that range from policy (Chapters 12 and 13) to planning (Chapter 14), and program levels (Chapters 15 and 16), but also raising issues of public involvement and negotiation at regional (Chapter 17) and local levels (Chapter 18).

The concluding Section VI reviews the value and interrelationships of the various chapters, putting them into perspective and relating to the future of SEA.

If this book catalyzes practitioners to combine economic and environmental considerations, to arm decision-makers before decisions are made, to inform theory by practice, it has achieved its goal. It is those practitioners to whom this book is dedicated.

REFERENCES

1. Sadler, B. and Verheem, R., *Strategic Environmental Assessment — status, challenges and future directions*, The Hague, Ministry of Housing, Spatial Planning and the Environment of The Netherlands, 1996.
2. Therivel, R. and Partidário, M.R., Eds., *The Practice of Strategic Environmental Assessment*, Earthscan, London, 1996.
3. Dalal-Clayton, B. and Sadler, B., *The Application of Strategic Environmental Assessment in Developing Countries*, draft report, IIED, London, 1998.
4. Therivel, R., Wilson, E., Thompson, S., Heaney, D., and Pritchard, D., *Strategic Environmental Assessment*, Earthscan, London, 1992.
5. Partidário, M.R., Strategic Environmental Assessment — principles and potential, in *Handbook on Environmental Impact Assessment*, Petts, J., Ed., Blackwell, London, 1999, chap. 4.
6. Clark, R., Making EIA count in decision-making, 17th Annual Meeting of the International Association for Impact Assessment, New Orleans, LA, 1997.
7. Partidário, M.R., Significance and the future of strategic environmental assessment, *International Workshop on Strategic Environmental Assessment*, Japan Environmental Agency, Tokyo, 1998, 13.
8. Verheem, R., Present status of SEA in The Netherlands, *International Workshop on Strategic Environmental Assessment*, Japan Environmental Agency, Tokyo, 1998, 48.
9. Partidário, M.R., *SEA — A Training Manual*, IAIA pre-meeting training course manual, unpublished.

Section II

SEA in a Sustainability Perspective

2 Making EIA Count in Decision-Making

Ray Clark

CONTENTS

INTRODUCTION

Twenty-eight years ago the United States passed landmark legislation that created the requirement to prepare a "detailed statement," for any "major federal action significantly affecting the quality of the human environment." When the U.S. Congress passed the National Environmental Policy Act (NEPA),[1] they envisioned that federal agencies would use the newly created tool, environmental impact assessment (EIA), to integrate environmental concerns with economic and social development. NEPA has been called the most powerful expression of national choice,[2] connecting a national purpose with a tool to achieve that purpose. Since then, thousands of analyses have been prepared for actions and projects. Since President Nixon signed NEPA in 1970, there have been nearly 30,000 environmental impact statements (EISs) produced in the United States. This does not include the approximately 50,000 environmental assessments (EA), a less detailed EIA, prepared annually.[3] Thousands more EIAs have been prepared worldwide by more than 100 countries that have adopted the EIA procedure.[3]

In the United States, EIA has been criticized for costing too much, being prepared after the decision is made, and engendering conflict and litigation. An early critic of EIA was Fairfax,[4] who argued that NEPA wrongfully assumes that there is perfect information traveling along a structured and disciplined path to one decision-maker who balances the environmental, economic, and technical considerations and optimizes resource use. Fairfax argued that decisions are made incrementally, with imperfect information, and it is costly and futile to use our scarce analytical and management resources in this manner. Others argue that decisions are divided among

1-56670-360-3/00/$0.00+$.50
© 2000 by CRC Press LLC

many individuals acting through a variety of social organizations, and no single individual can possibly interpret vast amounts of social and environmental data and use it for management purposes. Both friend and foe of the EIA requirement of NEPA are concerned about the extent to which it makes a difference in decision-making.

This chapter argues that while each of the criticisms has a "whisper of truth" to it, none of the criticisms are universally true. Adaptations to EIA can and should be tried as society strives to foresee environmental consequences and manage its future. The author believes that EIA is one of the great ideas of the 20th century and is a driving force in democratization of some nations, arguably the greatest political idea in human history. On the other hand, the author believes that EIA can be made more effective and more efficient, and contain less conflict by being prepared earlier in the decision-making process. New approaches should be developed that can help realize the full potential of environmental impact analysis.

Strategic environmental analysis is one new way of examining EIA by looking at when it is prepared, the scope of its analysis, and the level of detail required for an EIA. Advocates of SEA, however, must recognize that many decisions are really made incrementally, and that more often than not, decisions are made with imperfect information despite the effort and resources spent to gather data and information. Environmental assessment prepared at a truly strategic level will look different from traditional EIA while maintaining the principles of EIA. It will not replace traditional EIA; it will be an additional tool to help communities and countries develop sustainable paths.

This chapter is an attempt to recast the thinking of EIA in light of what we have learned over the past 25 years. Practitioners have learned that cumulative effects analysis is not another analysis, it is simply richer EIA. With Strategic Environmental Assessment (SEA), however, there are different principles involved than the manner in which EIA has evolved. What SEA brings to decision-making is early consideration (albeit broad) of the environment, incorporation of economic development and the material needs of human communities, early consultation with the public, and a consideration of alternatives before there is an irreversible commitment of resources. The U.S. has a long history of preparing "programmatic" EIA and many believe that this is the equivalent of SEA. In one sense, programmatic is "strategic," but realization of the full potential of SEA will require formulation of a new model of EIA and decision-making. In this model, strategic EIA will be prepared very early. It will be a short, concise analysis from which subsequent analyses will be tiered. It will be available to decision-makers before budgets are submitted, legislation is debated, and the site specific information is gathered. The analysis will focus on paths, not places. The second level analysis will focus on the programs the strategy yields. At a watershed or ecosystem level, multiple agencies operating within a common ecosystem will work together to prepare analyses on the initiatives within that ecosystem.

THE DECISION-MAKING MODEL

Initial decisions regarding U.S. policy or strategy are made by elected officials and appointed officials at the secretarial and assistant secretary level of federal departments. Political appointees make these decisions to put the stamp of their vision on society (or just an economic sector). Why else should one stand for election or choose the difficult path of public service? Their decisions are made in consultation with members of Congress and the staff within the legislative branch of government, often with the aid of the industries, associations, and public interest groups most affected by these decisions. Many major decisions are made at the departmental level. For instance, in the U.S.'s National Energy Strategy (1992),[7] the Secretray of Energy noted that over 90 of the strategy initiatives could be implemented without new legislation. While many of these initiatives were environmentally friendly, no NEPA analysis was completed for the strategy. The Department of Energy's reasoning was that the strategy did not contain any legislative initiatives because there are no specific proposals in the strategy and because no previous national energy plan triggered NEPA. This decision-making pyramid generally looks like Figure 2.1.

FIGURE 2.1 Strategic decision-making.

Many writers have concluded that the most effective way to improve the quality of impact assessment and influence the outcome of decisions is through political and administrative action. Andrews[6] makes the argument that one of NEPA's most fundamental limitations has been the rarity of its influence on truly major federal decisions at the policy, programmatic, and legislative levels. While the EIS is done at the project level, policies, legislative initiatives, or appropriations bills that underlie them are major federal actions that create far more pervasive impacts. Andrews cites government policies with perverse environmental results like agriculture crop-payment formulas, below-cost timber sales, fossil fuel and mining subsidies, and differential investments in highways as opposed to mass transit. A major finding of the U.S.'s study on NEPA's effectiveness[7] was that the process is triggered too late to become fully effective. NEPA's purpose to consider alternatives, weed out poor proposals, and support innovation is stifled by the timing of the analysis.

In 1994, the President's Council on Sustainable Development (PCSD)[8] concluded that environmental policy has been developed with too little regard for its economic or social consequences; economic policy is made with too little regard for environmental or social consequences; and social policy is made with too little regard for economic and environmental consequences. While the PCSD set up eight task forces to deal with integration of environmental, economic, and social issues, not one was established on how to use EIA as a tool to integrate these concerns into decision-making. Yet the vision laid out by PCSD incorporates many fundamental principles of EIA. Still searching for a path to sustainability, the Chairman of the Council on Environmental Quality committed to supporting base closure within the Department of Defense if the Department would agree to preparing an environmental sustainability analysis prior to recommending particular installations for closure or realignment (letter from Kathleen A. McGinty, Chairman of the Council on Environmental Quality, 1997).

The public clearly wants to be more involved in the sustainable development discussions. This requires being involved in decisions at the level when assistant secretaries and other high-level policy-makers are screening and choosing among options. While there is a true desire among many policy-makers to consult with the public, there is a belief among some that the public needs a concrete proposal to react to, rather than a mere conceptual notion. There is a reluctance to involve the public in amorphous discussions for fear of appearing ill-prepared, or worse, being sued. However, these proposals do get less quick with time — consultations among high-level policy-makers provide shape and foreclose options. Few of these strategic decisions get revisited. At this level, fewer people are involved and conflict is worked out among political elites.

Post Strategy. Of the 532 EISs prepared in 1994, about 77%* appear to have been prepared after someone had made the decision about strategic paths. Importantly, of the 125 EISs prepared by the Department of Transportation, only 3 were mass transit projects, 1 was a railroad, 12 were airports, and 102 were roads. (An additional 7 were prepared for bridge proposals.) More importantly, each of these EISs were prepared by different parts of the Department of Transportation, Federal Highway Administration, Federal Railway Administration, Federal Aviation Administration, and the Coast Guard.[9] All are separate administrative units reporting to the Secretary of Transportation.

In 1998, the U.S. Congress passed a $200 billion Transportation Efficiency Act for the 21st Century (TEA-21) and the Secretaries of Transportation in the 50 states are eagerly awaiting their share of the appropriations. By far, the largest percentage of the money will be spent on the nation's highway system. The TEA-21 was passed before such strategic questions as "how much should be spent expanding the highway system to relieve congestion" vs. "how much should go to preserving that which is already in place." While the executive and the legislative branches worked out the amount and agreed upon some earmarked projects, state officials will make most of the decisions about projects and will prepare the EIS required under NEPA. At this

---

* This includes draft, final, and supplemental environmental impact statements. The number of projects that are the subject of an EIS is lower.

level, there will be public involvement, controversy, and litigation over the effects of highway building on the ever-increasing sprawl. The funding formula that the U.S. uses to help the states favors large capital improvements, rather than maintenance of existing roads.

This bifurcation in decision-making is being noticed by the governmental institutions charged with reporting on the effectiveness of government. The General Accounting Office, an oversight arm of the U.S. Congress, recently issued a report on the Forest Service's decision making process and concluded that there are major inefficiencies in developing forest plans and reaching project level decisions. The study suggested that environmental analysis accompanying a plan or project be "tiered" or linked to a broader-scoped environmental study. Similarly, the Chief of Engineers Environmental Advisory Board in 1995 said that the "Corps [of Engineers] needs to review its integrated planning policy for all missions to ensure that EIA begins at the inception of a plan and continue through the completion of the project." One of the most vexing questions about EIA has always been how early is "early." Perhaps SEA can help answer that question.

WHAT IS SEA?

One of the most important and rapidly evolving trends in EA practice internationally is the recent progress with the application of EA to policies, plans, and programs, or SEA. One of the first hurdles facing the EIA profession is the very definition of SEA. Literature consistently says it is a systematic approach of addressing the environmental considerations and consequences of proposed policy, plan, and program initiatives. SEA is a decision-aiding process that can and should be applied flexibly to the decision cycle, recognizing that these terms mean different things and often cover different types of decision-making processes. In all cases, the approach taken should be consistent with EA principles. What does not automatically follow from this definition is the requirement to use EA procedures and methods in SEA, though their use and adoption for this purpose are widespread in some countries.

How is SEA related to decision-making? The scope and form of SEA will be contingent upon the function assigned, the policy and institutions that are in force, and the extent to which other comparable processes are used for similar purposes. SEA can be used to operate as a part of an integrated process or a separate approach that incorporates other factors (e.g., social and economic).

Strategic EIA has been an emerging topic within the EIA community for the past five to six years. The discussion often centers on the worth of SEA, and if worthy, at what point should SEA commence, in theory and in practice? In theory, Europeans require SEA at the policy, program, and plan level. In practice, some Europeans call an analysis done at the program level (a highway program) a strategic EIA, and some in the U.S. would agree with that characterization. Others require an EIA of budgets submitted to the legislature (Finland). A recent Austrian study on SEA concluded that Austrian plans and programs are only subject to a formalized EA to a very limited extent. Furthermore, the decision-making process is not explicitly influenced by the environmental components of planning strategies in any of the sectors' studies (agriculture, energy, forestry, industry, mining, etc.).[10] In some western countries, SEAs are required and shaped by mandatory procedures. In most countries, however, government agencies have applied SEA on a voluntary basis (Commission of the European Union). There is an argument made that Council on Environmental Quality (CEQ) regulations already require an environmental impact analysis at the policy and program level and that it is strategic environmental analysis by another name. In practice, it is rare that a policy level EIA is prepared. Even if one considers programmatic equivalents to SEA, only a small percentage of the 200 to 300 draft EISs prepared annually are programmatic (see Table 2.1 and Table 2.2).

TABLE 2.1
Draft (D) Draft Supplemental (DS), Final (F), and Final Supplemental (FS) Programmatic EISs 1983–96

Year	83	84	85	86	87	88	89	90	91	92	93	94	95	96
D	5	4	3	1	0	0	0	0	0	0	1	3	6	7
DS			2	2	1									
F	7	2	3	1	0	1	0	0	0	0	0	3	2	6
FS	2	0	0	2	0	0	0	0	0	0	0	0	0	0

TABLE 2.2
Legislative Environmental Impact Statements 1983–96

Year	83	84	85	86	87	88	89	90	91	92	93	94	95	96
D	0	0	0	3	0	1	1	2	1	1	1	5	5	2
F	0	0	0	0	1	1	0	0	0	1	0	1	1	0

Note: CEQ regulations specify that when only a "draft" will suffice to satisfy the requirements of NEPA. Wild and Scenic River designation requires a Legislative EIS and must have a draft and final LEIS.

A good example of a strategic EIS was not actually called "programmatic" or "strategic" It was filed as "Business Plan Environmental Impact Statement" by the Bonneville Power Administration (BPA).

The Commission of the European Union considers any analysis above the project level to be SEA. A 1994 study says that SEA alludes to "all levels of public decision-making which forgo the project level." In the European context, SEA allows for options that would "otherwise be out of scope such as alternate locations or routes."

It is likely that most EIAs prepared in Europe and the U.S. would consider alternate routes and locations. The European approach appears not so much "strategic" as programmatic. By the time an analyst is looking at "alternate routes or locations," many decisions have foreclosed options. This approach is entirely too late to discuss alternate means of providing transport or energy, frustrates the public, and has too little influence in the development of the nation. There are decisions about a highway that is appropriately made at the project level; the question, "Shall

---

BUSINESS PLAN FINAL ENVIRONMENTAL IMPACT STATEMENT

The electric utility market is increasingly competitive and dynamic. To participate successfully in this market and to continue to meet specific public service obligations as a federal agency, the BPA needs adaptive policies to guide marketing efforts and its other obligations such as its energy conservation and fish and wildlife responsibilities.

This EIS evaluates six alternatives to meet this need.

1. *No Action.* This alternative would maintain BPA's traditional activities in planning for long-term development of the regional power system, acquiring resources to meet customer loads, sharing costs and risks among its firm power customers and non-federal customers using the federal transmission system, and administering its fish and wildlife function, with the goal of fulfilling the requirements of the Northwest Power Act.

2. *BPA Exercises Market Influences to Support Regional Goals.* Under this alternative, in addition to its own activities to acquire energy resources and enhance fish and wildlife, BPA would exercise its position in regional power markets to promote compliance by its customers with the goals established by the Northwest Power Act.

3. *Market Driven BPA (The BPA Proposal).* BPA would change its programs to try to achieve its mission while competing in the deregulated electric power market. BPA would be a more active participant in the competitive market for power, transmission, and energy services, and use its success in those markets to ensure the financial strength necessary to fulfill its mandate under the Northwest Power Act.

4. *Maximize BPA's Financial Return.* Under this alternative, BPA would operate like a private, for-profit business. It would focus on limiting costs and investing its money where it can get the best return, while continuing to fulfill the requirements of the Northwest Power Act.

5. *Minimal BPA Marketing.* BPA would not acquire new power sources or plan to serve customers' load growth. Activities would focus on meeting revenue requirements through the long-term allocation of current federal system capability, while continuing to fulfill the other requirements of the Northwest Power Act.

6. *Short-Term Marketing.* BPA would emphasize short-term marketing (five years or less) of power and transmission power products and services, while continuing to fulfill the requirements of the Northwest Power Act From Business Plan Final Environmental Impact Statement (DOE/EIS-0183).

the highway traverse the wetland or bypass the wetland?" can be a strategic decision, but usually is not. Whether a highway crisscrosses a region is definitely a strategic decision. Using EIA to promote sustainability requires an assessment of whether to transport people and goods by highways or some other means. These are strategic choices.

Further evidence that the European model is similar to the U.S. model for "programmatic" EISs are the examples the European Union (EU) cites as susceptible to SEA:

1. land use plans
2. structure plans
3. siting or routing studies
4. technological programs
5. sectoral programs (e.g., energy sector)
6. policies

Number 3 is project-level decision-making, numbers 1, 2, 4, and 5 are programmatic in nature, numbers 4 and 5 *can* be strategic if the question is framed broadly enough, and number 6 is the only level that is inherently strategic.

CHALLENGES TO IMPLEMENTING SEA IN THE UNITED STATES

While SEA is a promising avenue for incorporating environmental considerations into the highest levels of development decision-making, it is still at a relatively early, formative stage. Many practical questions remain about procedures, methods, and institutional frameworks. Policy-makers are being urged to make decisions at a larger and larger scale, indeed even at a global scale. While EIA practitioners recognize the phenomenon, the profession has not been very successful in adapting EIA to the enormous task. There are several major challenges that must be resolved to make SEA attractive to policy-makers.

1. *Definition.* In any new approach, the first problem is to define exactly what the notions are and get a general understanding of the concept. Therivel, Partidário, and others have moved the profession a long way toward a common definition of SEA. Some remain unconvinced, however, that EIA of one highway or one nuclear power plant is strategic EIA. If policy-makers are grappling with sustainablity strategies, analysts must provide data and information about energy paths and transportation direction. This chapter argues that strategies yield policies and programs, and policies and programs yield multiple projects in multiple sectors (energy, transportation, etc.) in specific ecosystems or watersheds. Within those ecosystems, individual power plants or roads are cited. To be attractive to decision-makers in the U.S., SEA must be different than programmatic EISs. One major reason for this is that data show that programmatic EISs

take an average of about five additional months to complete and are about four times as expensive.[11]

2. *Organizations.* Current organizations, at least within the U.S., are not cohesive enough to work at a strategic level within one sector. Transportation projects are proposed by the individual states. The legislative branch (particularly the transportation committees) and the executive branch (particularly the Secretary of Transportation) make incremental decisions regarding whether or not to pursue these proposals. Transportation decisions are collaboratively made among the individual states, the legislative branch, and the executive branch (the president and his cabinet). It is important to note that the U.S. Department of Transportation (DOT) is divided into six administrations (or agencies) and the St. Lawrence Seaway Department Corporation,* each dealing with a different mode of transportation and each with its own administrative head appointed by the president. While many transportation decisions are made at the departmental level, the EIA is prepared at the agency level, usually in the field. A source of management pride is that implementation of the programs are done at the field level, and the field staff are given wide autonomy in deciding how to accomplish a project, but they are given little discretion in deciding whether to implement a program. (Oddly enough, the states who originally propose a project, later become the implementers.) Each administrator within the DOT develops regulations on how the NEPA analysis will be prepared, including what actions are categorically excluded from EIA, when the public will be involved and to what extent, and the breadth that alternatives will be considered. While there is an office of intermodalism in the Secretary of Transportation's office, that is not where NEPA operates. One may conclude that the obvious solution to this bifurcation is to combine and integrate DOT requirements. That oversimplifies the real organizational problem. The natural environment doesn't really respond to the departments. It is more important for the departmental level to understand environmental impacts of different transportation paths at a broad level across agencies (and think in terms of cumulative effects) within a particular ecosystem. It is at an ecosystem level where it is likely that most of the species and environmental resources are more affected by the combination of different federal and non-federal activities rather than the combination of transportation projects scattered over various ecosystems from the northern high plains to the southern Appalachians.

3. *Data.* Data and information feed analysis. Analysts rarely have all data leaders believe they need for decision-making. National and international efforts are only beginning to collect much-needed, long-overdue, environmental baseline data. Even with existing data, federal and state agencies do not have the infrastructure to coordinate sharing data or information. At a strategic level, there are even less data, interpretation is foggier, and

---

* Federal Highway Administration, Federal Aviation Administration, Federal Railroad Administration, Maritime Administration, Federal Transit Administration, U.S. Coast Guard.

acceptance by decision-makers is less certain. It is difficult for an analyst or a policy-maker to assess the environmental effects of a conceptual idea absent of setting and time. There is always the question of how much information a policy-maker needs to be within a comfortable decision-making space. Often, it may be sufficient to judge the consequences on several key macro issues: (1) the use of natural resources, including energy and raw materials; (2) the quantity and quality of waste streams; (3) emissions to air, water, soil; (4) human health and safety; and (5) use of space.[12] The higher level the decision is, the less likely an environmental analyst will be able to quantify impacts or even predict the probability of some impacts. Still, while SEA can provide a level of environmental information at a macro level, it must also assure decision-makers and the public that while such uncertainties exist, it is appropriate to move to the next stage of decision-making.

4. *Uncertainty.* The lack of data is often cited as a reason EIA cannot be prepared at a higher level. Analysts crave data (though there are times when they are even entangled in it). There always will be some people adverse to risk unwilling to make decisions or allow decisions to be made without virtual certainty. Indeed, there are points in the decision-making stages when more detail (like engineering drawings) is required. Usually, however, that is at a tier (or two) following the strategic decision. It is not *how* you are going to do something, it is *whether* and *what* you are going to do that requires strategies. Nations make multimillion dollar economic decisions about whether to build highways and with enough certainty to believe the wisdom of that choice. They proceed in view of the uncertainty about its technical feasibility and its economic viability. Nations take risks to move economies forward. This is the appropriate time to have the EIA practitioners display the tools (scenario development, risk assessment, etc.) to develop environmentally sustainable paths that allow decision-makers to calculate the environmental and social dimension of their decisions. As the EIA profession gets more sophisticated at this business, EIA will incorporate monitoring and adapting strategies, rather than insist upon upfront certainty.

5. *Litigation.* In the U.S., NEPA has been used as a tool to stop federal projects, to alter projects, to avoid environmental impacts, and to establish a body of case law that recognized the environment was a paramount concern in agency decision-making. While the courts have said that NEPA creates a procedural obligation on government agencies to consider the environment, the Supreme Court (the third co-equal branch of government) has instructed lower courts not to substitute its judgements for that of the federal agency. As long as the agency has taken a "hard look" at the environmental consequences, they will likely prevail in court. In 1979, the court ruled in Andrus vs. Sierra Club* that an EIS is not required as part of the budget process, thereby closing an early opportunity to think

---

* Andrus vs. Sierra Club, 42 U.S. 347 1979.

strategically about environmental obligations. Although the court observed that if environmental concerns are not interwoven into the fabric of agency planning, the action-forcing characteristics would be lost. The court played the major role in shutting off such analysis. Federal agencies have learned the procedural parts of NEPA very well. They prepare EISs that document every conceivable impact, develop an administrative record and provide the public an opportunity to comment. This often avoids litigation. (It does not necessarily avoid adverse publicity.) Strategic EIA, as this author defines it, has many pitfalls and decision-makers are reluctant to embrace it now. The lack of a concrete proposal, or "mere speculation," is already explicitly exempted from NEPA review through case law. Voluntary exposure to potential litigation would be treated by staff and agency leadership as unnecessary. What may attract SEA to U.S. policy-makers is the opportunity to use the benefits of EIA without the procedural traps, particularly in litigation. A major benefit of SEA is the open discussion of the future, a clear goal of the framers. Litigation should not drive policy. There is, after all, no irreversible commitment of resources at the strategic decision-making level and that may be the largest benefit of SEA over project-level EIA. SEA should be exempt from litigation. There will be ample time for litigation at the project or program level should the need arise, and this author is not suggesting that EIA should not be challenged by litigants.

6. *The Problem of No "Proposal."* Is spending resources toward the development of a new generation of nuclear power plants a "proposal"? There is no location attached to it. Even so, it is possible to assess the difference among alternate strategies. Even at a macro level, it is possible to assess which have more land occupation, which results in more habitat loss, whether one strategy results in more emissions of air pollutants, whether one approach consumes more energy, and whether one path may have more acceptable risks. It is possible to determine the public's willingness to accept risks. Designers and the strategists unwittingly transfer enormous conflict to communities without ever knowing what site may be chosen.

7. *Capacity, Knowledge, and Skills.* Preparing EIA at the project level requires advanced skills and the pool of qualified professionals is low. SEA is another rung up the ladder of complexity and closer to the policy level decision-makers with high expectations. EIA practitioners haven't yet mastered strategic analyses like carrying capacity and sustainability thresholds. This new approach requires analysts who understand not only EIA, but the business sector being studied. With some successes, (seen through the eyes of policy-makers), the profession can advance this part of EIA. With several highly visible failures, SEA will set back, perhaps diminishing the opportunity to further develop the idea. Along with these analytical skills must come an ability to bring closure to the process at a cost that seems reasonable.

8. *Political Will*. Policy-makers do not take unnecessary risks, but will take risks that they can manage. They seek as reward recognition for breaking new ground in the advancement of the organization's core goals. The entire sustainable development movement has brought EIA practitioners an opportunity to help top-level decision-makers use the EIA tool to successfully set a sustainable course. Practitioners know that EIA can be a tool that turns rhetoric into action; many policy-makers do not yet believe that.

9. *Role of the Public*. How early the public should be involved in EIA remains one of the most vexing questions. The public's acceptance to play at a stage when things are uncertain, when there is no proposal, and agree not to sue because the policy-maker is proposing an ill-formed, half-baked idea is critical to the success of SEA in the U.S. It is generally true that the public does not want to participate in an EIA until there is a concrete proposal. Participation in EIA is extracurricular for most of the public. After preparing dinner, church services, soccer league, and all the other demands on the time of families, EIA practitioners seek their time and attention. When the public gives their time, policy-makers should see EIA and public involvement as a useful and helpful product, not a therapeutic process for people devoted to procedure.

10. *Integration*. There has always been an argument within the EIA community in the U.S. regarding whether EIA is an objective, analytical tool, or whether it is an integrative, planning tool. That is, whether EIA integrates economic, social, and environmental concerns into one analysis for decision-makers. Currently, EIA in the U.S. limits its scope to the biophysical impacts of a decision. This is not very useful for decisions at a strategic level because so many other factors are important. SEA has the opportunity to combine economic development, environmental protection, and community well-being in one analysis, not three.

CONCLUSION

Policy development is a dynamic process and it is inevitable that policy issues will never be as precise as the EIA profession has assumed them to be. Because decisions are incremental and there is not always a precise stage at which government decides public policy, so too should EIA evolve to a tiered, incremental approach. SEA moves the focus from one place, one site, to a more strategic level so that policy-makers can see how their entire operation fits in a national or even global context. Clearly by the time a project is assessed, there are few decisions left to be made about whether a proposal will proceed. This is appropriate for some types of projects, but also for use in sustainable development strategies. Do we need different procedures for SEA than EIA? Yes. While the principal analytical elements are similar, there are significant differences. SEA must be more flexible, allowing the decision-maker to take those elements that are useful. Inception and extent of public involvement will be up to the decision-maker who must see the benefit of SEA. SEA is significantly more complex than EIA and it will require developing a professional capacity to ensure it success and acceptance. It yields an iterative, continuous look at the environmental effects of an action. This emphasis on monitoring, rather than certainty, will yield new information that under ordinary circumstances may procedurally call for a new EIA. Because SEA should lead to a shorter, simpler, more open process, perhaps SEA itself should have no procedural requirements. The new model does it a different time, covering a different scope, maintaining the science and art of IA while modifying those things that remove the obstacles of doing it earlier.

However, one should not oversell SEA as an analysis that can predict sustainability. SEA can help light many different paths, but it cannot prevent stumbles. Importantly, SEA requires the integration of environment, economic, and social analysis. Perhaps one of the important first steps is to train decision-makers about the impact assessment toolbox (and the limitations of the toolbox). There remain many complexities, but the world needs devices, tools, and instruments to tell which way the wind is blowing, even if that instrument is just a wet thumb.

REFERENCES

1. National Environmental Policy Act, 42 United States Code Sections 4321-4370.
2. Blaug, E.A., Use of environmental assessments by federal agencies in NEPA implementation, *Environ. Profession.*, 15, 57-65, 1993.
3. Wood, C., What has NEPA wrough abroad?, in *Environmental Policy and NEPA*, Clark, R. and Cantor, L., Eds., St. Lucie Press, Boca Raton, FL, 1997.
4. Fairfax, S.K., A disaster in the environmental movement, *Science*, 199(917), 743–748, 1978.
5. U.S. Department of Energy, *National Energy Strategy*, Washington, D.C., 1992.
6. Andrews, R.N.L., The unfinished business of national environmental policy, in *Environmental Policy and NEPA*, Clark, R. and Canter, L., Eds., St. Lucie Press, Boca Raton, FL, 1997.
7. Council on Environmental Quality, *The National Environmental Policy Act: A Study of its Effectiveness After Twenty-five Years*, Executive Office of the President, Washington, D.C., 1997.
8. President's Council on Sustainable Development, *"Sustainable Development: The Challenges and Opportunities,"* Executive Office of the President, Washington, D.C., 1994.
9. Council on Environmental Quality, *Environmental Quality*, Executive Office of the President, Washington, D.C., 1995.
10. Lee, N., Barker, A.J., Wood, C., and Jones, C.E., Eds., EIA Newsletter 12, University of Manchester, Manchester, U.K., 1996.
11. U.S. Department of Energy, *Lessons Learned*, Quarterly Report of the NEPA Policy Office, Washington, D.C., 1998.
12. Verheem, R., *SEA of the Dutch Ten Year Progamme on Waste Management*, Dutch EIA Commission, June 10, 1994.

3 Strategic Sustainability Appraisal — One Way of Using SEA in the Move Toward Sustainability

Maria Rosário Partidário and Filipe V. Moura

CONTENTS

INTRODUCTION

Development policies established in the post-Rio era generally refer to sustainability as its ultimate goal. This is quite understandable given the universal acceptance of the United Nations Conference for Environment and Development recommendations, in particularly, Agenda 21, where sustainability is adopted as the main principle. However, the particular definition of sustainability, and its practical implementation, is still a major handicap. It is very hard to establish what is sustainable and how sustainability can be measured. The only way to find that out is by trial-and-error approaches and by applying early and in the best possible way.

1-56670-360-3/00/$0.00+$.50

Sustainability indicators are increasingly viewed as quite instrumental in the process of giving a certain "dimension" to sustainability. Although normally qualitative, indicators can be better than a vague notion. This chapter contains material related to research conducted with the purpose of developing a concept that, by using sustainability indicators, relates Strategic Environmental Assessment (SEA) to sustainability objectives and targets, through the adoption of a simple approach: Strategic Sustainability Appraisal.

Strategic Sustainability Appraisal (SSA) is not a procedure like SEA or EIA (Environmental Impact Assessment). Rather, it intends to act as an instrument that, while operating in the context of an SEA procedure, provides a framework based on quantifiable approaches (thresholds and targets) and a mechanism to check on the sustainability trend of a proposed, or on-going, strategy (whether policy, planning or program). In this context, indicators of sustainability are used to relate policy objectives to sustainability, while thresholds and targets enable the quantifiable measure of the strategy effectiveness in achieving sustainability.

SSA was initially developed as an attempt to address sustainability in a more operational way in the context of the SEA baseline methodology of COMMUTE, a European Commission transport research project.[1] By providing a framework that integrates the results of other assessment tools into an overall approach, the purpose of SSA was to ensure that issues of sustainability have a specific role-play in the planning and decision-making process. SSA attempts to achieve this objective by highlighting and articulating particular sustainability objectives, priorities, and targets in specific decision systems.

The following sections present SSA's rationale and concept, particularly that of indicators of sustainability, its relationship with traditional indicators (e.g., climate change, acidification, noise, air pollution, water pollution, etc.), and some key procedural principles and assumptions that will ensure the sustainability approach in a planning and decision-making process. While SSA can potentially be used in a variety of contexts, its application can be better understood if it is described with respect to a concrete case study. This chapter will highlight the application of SSA to the transportation policy and sector.

RATIONALE FOR STRATEGIC SUSTAINABILITY APPRAISAL

POLICY CONTEXT

The Expert Group on the Urban Environment of the European Commission[2] have argued that "the principle of integration is never more important than for combining the goals of sustainability with the realities of urban management". This can be taken as an accurate statement for sustainability in transportation systems as well. On a strategic level, it is necessary to elaborate on a comprehensive and broadranging vision of what sustainable transportation systems may be. This will represent the framework within which actions can be formulated and implemented, through practical incremental steps, guided by the general goals in a strategic framework.

Agenda 21 addresses sustainability in transports in various different sections. In essence, sustainability in transportation systems is understood as:

- cost-effective, more efficient, less polluting, and safer transportation systems than existing ones
- integrated land-use and transportation planning strategies with a view to reducing the environmental impacts of transports

The European Union (EU) Common Transport Policy also adopts sustainability as a key strategic goal in the context of the recommendations of Agenda 21, referring particularly to the need to avoid depletion of natural resources and avoidance of environmental pollution.

One important issue about sustainability in Agenda 21 is the need to ensure that the adequate information is provided to decision-makers. This is where environmental assessment (EA), and particularly SEA, can play a key role. By associating to SEA the development and promotion of the global use of indicators of sustainable development, this can provide a solid basis for decision-making at all levels, and contribute to a self-regulating sustainability framework of integrated environment and development systems.

RELEVANCE TO THE SUSTAINABLE PLANNING AND DECISION-MAKING PROCESS

There is a world general rationale for the adoption of sustainability approaches within the planning and decision-making processes on which the above policy references provide a minimum example. It is commonly consensual that policy objectives and targets must be established to provide a framework for the measurement of adopted indicators.[3] Table 3.1 offers examples of indicators of Sustainable Development (ISD) and targets developed by several international organizations.

For SSA to be implemented in the planning and decision-making process, it is necessary to ensure that:

1. Sustainability criteria for decision-making is established and included in key decision-making points.[12]
2. Clear sustainability targets are established that satisfy sustainability criteria.
3. Flexible mechanisms are introduced to ensure that before planning and decision-making takes place, policy-makers and other stakeholders adapt the sustainability criteria and targets to their correspondent level of decision (local, regional/state, national).
4. Mechanisms are introduced to highlight situations that go beyond the thresholds defined by the sustainability criteria and targets (red flag).

Two types of indicators can be used for this purpose:

TABLE 3.1
Examples of Indicators and Targets for Sustainable Development

Theme	Indicator	Targets
Climatic Change	Quantity of CO_2-emissions	Stabilisation in the year 2000 on the level of 1990's CO_2 emissions (progressive emission reduction until 2005 and 2010) 3300 millions of tons per year[4]
		For the year 2000, a threshold is proposed at 180 megatons of greenhouse gas emission[5]
Acidification	Quantity of SO_2-emissions	Until 1994 stabilisation at the EU-level of 1990's SO_2 emissions. Until 2000 reduction of 30% (approximately 13 millions of tons per year)[4]
		For the year 2000, a threshold is proposed at 4000 acidification equivalents per ha[5]
	Quantity of NO_x-emissions	Until 2000 reduction of 35% on the level of 1985's NO_x emissions (approximately 9 millions of tons per year)[4]
Air Quality	Emissions of: VOC, SO_2, NO_2, particulates and black smoke, CO, lead, Cd and other heavy metals, organic compounds	Until 1996 (1999) reduction of VOC anthropogeneous emissions of 10% (30%) on the EU-level of 1990 (approximately 13 millions of tons per year (10 millions of tons per year))[4]
		Different targets according to different situations. Heavy Metals: at least 70% reduction from all pathways of Cd, Hg and Pb emissions in 1995[4]
		Application of existing regulations on SO_2, NO_2, Pb, particulates and black smoke[4]
	Emissions of greenhouse gases	Stabilization of GHG concentrations in the atmosphere at a level that would prevent dangerous anthropogenic interference with the climate system[6]
	Emissions of sulphur oxides	Reduce sulphur emissions by 30% from 1980 levels by 1993[6]
	Emissions of nitrogen oxides	Reduce nitrogen emissions by 1987 levels by 1995[6]
	Consumption of ozone depleting substances (ODS)	Complete phaseout of ODS[6]
	Ambient concentrations of pollutants in urban areas	WHO levels
Coastal Zones	Coastal zone as a percent of total area	Each country's coastal zones should not be depleted by more then 10% f the total area[6]
Soil Quality	Land affected by desertification	Reduce the area and % of land affected by desertification and/or reduce the severity of desertification[6]

TABLE 3.1 (CONTINUED)
Examples of Indicators and Targets for Sustainable Development

Theme	Indicator	Targets
Nature and Biodiversity	Protected area as a percent of total area	According to the Caracas Action Plan of the World Parks Congress, the established target for preservation of biodiversity and nature was of protecting at least 10% of each of the world´s major bioma. At an european scale, this goal would generically correspond to a protection of at least 10% of dry forests, moist forest, grassland, scrub, marsh, wetlands and mangroves[5] Enlargement of the existing protected areas, according to the EU Natura 2000's programmes, with their respective maintainance and preservation[4]
	Threatened species as a percent of total native species	Threatened species are less than 1% of the total species in any class[6]
	Protected forest area as a percent of total forest area	IUCN guidelines
Water Quality	Groundwater quality	The EU guide levels are of 5.6 mg/l for nitrogen (NO_3-N) concentrations with a maximum concentration of 11,3 mg/l in groundwater for drinking purposes, 170 kg/ha as limit on use of nitrogen in artificial fertilisers or manure, and 0,1 µ/l for individual substances and 0,5 µ/l for total pesticides as maximum concentrations of pesticides in groundwater[4]
	Eutrophication processes	In the Dutch's NEPP, the sustainability limit for eutrophication, according to their index of eutrophication[5]
	Domestic consumption of water per capita	At least 40 litres per capita per day of safe water in urban areas by the year 2000[7]
	Concentration of faecal coliform in freshwater	WHO drinking water thresholds, adopted by most countries
	Waste-water treatment coverage	Quantitative and qualitative discharge standards for municipal and industrial effluents are applied by the year 2000[6]
	Density of hydrological networks	World Meteorological Organization (WMO) Standards
Waste Management	Generation of hazardous wastes	Prevent or minimize the generation of hazardous waste as part of an overall integrated cleaner production approach[7]
	Imports and exports of hazardous wastes	Total ban of transboundary movements of hazardous wastes[6]

TABLE 3.1 (CONTINUED)
Examples of Indicators and Targets for Sustainable Development

Theme	Indicator	Targets
	Generation of radioactive wastes	Reduction of radioactive wastes generation[6]
Noise	Population within the 65 dB(A)-isophon (Leq) during night-time	Population should not be exposed to noise levels higher than 65 dB(A) during night-time
	Population within the 85dB(A)-isophon (Leq)	Noise level should never exceed a level of 85dB(A)
	Population within isophons greater than 55 and 65 dB(A) (Leq)	Proportion of the population already exposed to noise levels between 55 and 65 dB(A) should not suffer any increase
	Population within isophons greater than 55 dB(A) (Leq)	Proportion of the population already exposed to noise levels below 55 dB(A) should not suffer any increase above that level
Spatial impact	Number of (different types of) settlements cut through, weighted with the number of inhabitants and possibly with the type of infrastructure	Maintain actual levels of soil occupation by transport infrastructure, which corresponds approximately to 3% of Europe's territory[8]

1. traditional indicators, i.e., quality-of-life and environmental indicators, which may be assessed in a sustainability framework
2. indicators of sustainability, referring to clearly identifiable targets or thresholds of sustainability

CONCEPTS

STRATEGIC SUSTAINABILITY APPRAISAL

Strategic Sustainability Appraisal (SSA) can be understood as an integrative approach, hosted in an SEA framework, which bears on sustainability priorities and criteria defined at the policy level, and translates these into measurable indicators. In this way, SSA will act as a mechanism that aims to demonstrate the extent to which sustainability is met, thus offering a common framework for the appraisal of sustainability efforts.

As such, the key issue of SSA is not to discuss which indicators it may measure, but instead, the range of values indicators may adopt within a sustainability policy context, and the threshold point beyond which a sustainability situation can no longer happen or be ensured.

SSA is therefore a tool designed to contain a set of targets for indicators of sustainable development. It is a proposal for the adequate combination, or aggregation, of such indicators, and the benchmarking for appraising the success in achieving sustainability.

In developing strategic sustainability appraisal, the concept of carrying capacity is absolutely crucial. The Dobris Assessment[9] refers to the capacity of natural systems to provide energy and materials, and to absorb the interference of pollution and waste as the critical threshold within which the world economy can expand. But regarding sustainability, it also raises the question: how much human-induced change can the global environment sustain? Defining carrying capacity is therefore critical to sustainability appraisal.

> For the human population, the carrying capacity is the maximum rate of resource consumption, waste and emissions production that can be sustained in the long term globally and regionally without impairing ecological integrity and productivity. In order to be sustainable, the level and rate of natural resources depletion and pollution emissions should be no greater or faster than the level and rate of regeneration or environmental absorption.[9]

Carrying capacity and sustainability are complementary concepts that share a common difficulty: they are difficult to quantify. It is possible that there will never exist one single measure to define it. But in an environmental resources management context, they are indispensable ingredients of sound development. By adopting and integrating an approach such as Strategic Sustainability Appraisal, supported by sustainability thresholds and indicators, planning and decision-making processes will be contributing to an incremental, and iterative approach, to more sustainable practices.

Indicators of Sustainable Development (ISD)

"Sustainability indicators are definable, measurable features of the world whose absolute levels or rate and direction of change are intended to reveal whether the world (or a city) is becoming more or less sustainable."[9]

"ISD can be seen as a measurement (or series of measurements) which can be used reliably as a signal to specify the current state, or monitor the change, of factors having an impact on sustainable development."[10]

In a policy context, indicators emerge in a two-way process: they are implied or defined by policy objectives, but they also help to define and mold policy aims. The choice of an indicator is therefore not purely a matter of technical need, but inescapably a matter of policy choice with significant consequences.

In this context, indicators of sustainability play a fundamental role in constituting notions of sustainability or of what is sustainable development. However, measurement capacity and policy significance are both essential requirements and are not always compatible. The success with sustainable indicators lies in the best achievable combination.

A number of functions for indicators of sustainable development have been identified in the context of the POSSUM project:[11]

   - To identify sectors or impacts which are having an adverse effect on sustainable development;

- To measure the extent to which policies are achieving objectives of sustainable development;
- To simplify and communicate a large amount of data using a smaller amount of representative, meaningful information.

This last point is particularly important for communication with different stakeholders. Also, in order to mobilize the public, these tools should be developed, keeping in mind areas of public concern, enhancing the potential for public participation.

When referring to the identification and selection of potential ISD, two approaches have been suggested, whereby indicators are derived from principles, goals, sectors, issues, and/or causal relationships. These two approaches are described in the POSSUM project report (1997) as follows:

Top-down approach

The "top-down" approach is characterized by a deductive, comprehensive, systematic strategy using a combination of goal-based, domain-based, and issue-based frameworks. The goal-based frameworks start with the goals of sustainable development, and indicators are formulated around each goal. The other framework approaches refer to indicators established for environment, economy, and society domains (domain-based frameworks) and to "traditional" indicators defined for issues more readily understandable and fairly simple to construct (issue-based frameworks).

Bottom-up approach

This approach is characterized by an inductive, knowledge-based strategy. It involves the review of existing proposed and used indicators of sustainable development, as available in the literature, and the selection of relevant indicators.

For example, ISD are used in the POSSUM project as a means of identifying policy targets and assessing the impacts of policy packages across a range of key sustainability issues comprising the three themes of regional development, economic efficiency, and environmental protection.

In April 1995, on occasion of its third session, the Commission on Sustainable Development (CSD) approved a work program on ISD. The work program included a list of 132 indicators organized in the Driving Force, State, Response Framework. In this framework:

1. *Driving Force* indicators represent human activities, processes, and patterns that impact on sustainable development.
2. *State indicators* outline the "state" of sustainable development.
3. *Response indicators* illustrate policy options and other responses to changes in the state of sustainable development.[6]

The purpose of this work program is to provide users at the national level with sufficient information about the concept, significance, measurement, and data sources for each identified indicator so as to facilitate data collection and analysis.

Indicators are chosen according to their relevance to national priorities, goals, and targets.

The process was coordinated by the United Nations Department for Policy Coordination and Sustainable Development (DPCSD), but builds upon indicator work being carried out in several organizations.* Table 3.2 indicates the number of indicators on social, economic, environmental, and institutional issues according to the classification of Driving Force – State – Response. In the appendix, a short-cut of the list of ISD is presented.

TABLE 3.2
Grouping of ISD

	Driving Force	State	Response
Social	14	14	13
Economic	6	12	8
Environmental	21	17	16
Institutional	0	8	1

The ISD on environmental issues is clearly the largest. However, as indicated in the table in the annex, the practical application of these indicators is not yet feasible, in part due to the lack of available data and missing targets.

TARGETS

Defining quantifiable targets is often more difficult than identifying indicators. Targets are normally political, cultural, and geographically specific. They must relate to the scale of decision-making, whether a regional, or national scale. Clearly, the definition of targets is fundamental in working with ISD, as recommended by the UK Round Table: "ISD should, if possible, have targets attached."

Where targets have been established (e.g., United Nations Organization (UNO), World Resources Institute, EU 5th Environmental Action Programme, Dobris Assessment [Europe's Environment], Dutch National Environmental Policy Plan), they can be taken as benchmarks. However, there is still a long way to go before an acceptable common set of targets for ISD are identified and established (Table 3.3 indicates the proportion of the United Nations ISD, with established targets).

---

* Organizations which have contributed both to the development of the indicators and to the preparation of the report include the following: the United Nations Department for Policy Coordination and Sustainable Development (DPCSD); the United Nations Children's Fund (UNICEF); the United Nations Conference on Trade and Development (UNCTAD); the United Nations Environment Programme (UNEP); the Food and Agriculture Organization of the United Nations (FAO); the United Nations Educational, Scientific and Cultural Organization (UNESCO); the World Health Organization (WHO); the World Meteorological Organization (WMO); the World Bank; the Organization for Economic Co-operation and Development (OECD); the International Conservation Union (IUCN); the International Institute for Sustainable Development (IISD); and the World Wide Fund for Nature (WWF) among others.

TABLE 3.3
Proportion of the United Nations ISD
with Target Attached

	ISD with an Identifiable Target
Social	49%
Economic	0, 4%
Environmental	53%
Institutional	33%

Where no targets exist, they should be specified on a trial-and-error basis to support decisions in a sustainable way. This trial-and-error approach should be essentially oriented by the established policy goals of the surveyed system. If targets are to be set, then the rationale behind them should be made clear, and those with responsibility for achieving these targets should if possible, be identified. Notice that a trend in what is perceived to be a desirable direction can be a valid target in its own right, even without a stated-end-point.

Table 3.1 synthesizes a number of ISD and related targets as collected from the literature. It is presented here to illustrate the relationship between ISD and targets. The following sections will bear on these concepts.

GENERAL MECHANISM FOR IMPLEMENTATION

This section refers to the applicability of SSA. It implies "measuring" the ISD, not as a direct quantification, but as the establishment of the thresholds which the indicators, measured with any other measurement tool (e.g., cost-benefit analysis, multi-criteria analysis, geographic information systems, or life-cycle assessment) should meet in a sustainable context. The difference between the real value of the indicator, measured with any of the measurement tools, and the sustainable value, represent the effort that must be developed at the technological or policy level to achieve sustainability.

Analytically, this process can be represented as:

- *SSA*: Indicator A - Y (threshold)
- *Other measurement tool*: Indicator A - X (real value)

a) If $X < Y$, then the situation is below threshold and is acceptable, foreclosing sustainability.
b) If $X = Y$, then the situation is critical. Efforts must be developed to re-establish conditions for sustainability.
c) If $X > Y$, then the situation is non-sustainable, since the threshold has been overtaken. Efforts must be developed to re-establish conditions for sustainability.

In the following section, an indication is also given of the kind of sustainability indicators that may be adopted, and the need to establish intermediate indicators that will enable the achievement and use of the proposed sustainable indicators.

The key element in this process, again, is the identification of policy criteria and targets that may be used as sources for "measurement" references in a sustainability context.

APPLICABILITY OF INDICATORS IN THE SSA CONTEXT

This section takes stock on the aplicability of the indicators in a sustainability perspective. A key element is the identification of sustainability targets through policy. The 5th Environmental Action Programme (5th EAP), and its recent review, provide a set of environmental themes, as well as related targets with a timely horizon for the year 2000. Some of the themes covered by the 5th EAP and its review are: climatic change, acidification, nature and biodiversity, air quality, water quality, coastal zones, soil quality, and noise.

Traditional indicators, especially environmental indicators, are suitable for SSA if thresholds of sustainability are previously determined and if they are applied in each national context or, globally, depending on the approach scale. After collecting and integrating the existing information on sustainable targets and thresholds, lacking targets should be pointed out (e.g., coastal zones and soil quality, for which no targets have been identified). Further research should be performed to determine these targets and to identify new and more integrating sustainability indicators.

After collecting and integrating existing information on sustainable targets and thresholds, targets that may still be lacking identification or quantification should be indentified. And, as referred to in Deliverable 1 of the POSSUM project,[11] "if thresholds and targets cannot be determined, the desirable trend direction should be stated."

THE ROLE OF SSA IN THE CONTEXT OF SEA

SSA can be understood as an integrative approach that aims to translate sustainability priorities and criteria into measurable indicators. For both sets of indicators (traditional and sustainability ones), not only do measures and objectives have to be defined, but also concrete sustainability thresholds and ways of determining and upgrading them. Sources of information, methods, models and functions for measurement of traditional indicators are those currently used (e.g., the *CORINAIR** model in the *Air Quality* theme for modeling air pollutants dispersion). In regard to indicators of sustainability, such tools have to be identified and/or developed in order to enhance its implementation.

The following is a suggestion for key points in making the role of SSA in the context of SEA more concrete:

---

* European program for inventory of air emissions.

1. Introduction/scope definition: identification of adopted national/regional sustainable targets according to the policy, plan, or program's objectives and alternatives

   Where no national/regional sustainable targets exist, targets should be defined and used as quality standards. This should happen at the earliest possible stage of the undertaking SEA. A library of international standards (sustainability thresholds) could be made available.

2. For every indicator allow for three kinds of thresholds:

   I. Sustainability (The Target)

   II. Achievable Threshold (Quality Standards)

   III. Critical Level (under or above which strong impact is expected)

3. The role of SSA could be schematized as follows (Figure 3.1):

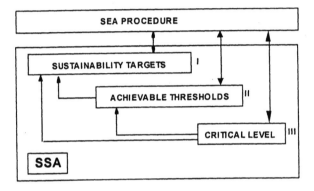

FIGURE 3.1   The role of SSA in the context of SEA.

The main principles on which SSA should be based are

I. should be the final target

II. should tend to I, while corresponding to quality standards

III. should show "red flag" beyond, or below, which target levels are not acceptable (strong impact), implying that earlier steps and premises should be readdressed

To demonstrate the application of this approach, the following example can be considered. Values included are not necessarily realistic.

Example

Type of indicator: Response
Issue: Accessibility
Indicator: Trip length for daily commute
       Thresholds I: 15 min (sustainable target)
                  II: 30 min (achievable threshold)
                  III: 60 min (critical level)

If when running, the model Threshold III is not achieved, that means that the situation is likely to determine a significant impact and the model premises need to be readdressed.

If when running, the value of the model Threshold II is achieved, or remains between II and III, that means that policy measures (in the medium to long-term) regarding technology innovations, different land-use or activities distributions, housing, employment, etc., should be reviewed so that policy conditions that provide rationale for transport policies are changed.

CONCLUDING REMARKS

Strategic Sustainability Appraisal represents an embrionic form of a potential mechanism to ensure the integration of sustainability issues, concerns, and priorities into the decision-making process. An attempt to demonstrate its interest was made in the transports case presented.

Comments received regarding an earlier version of this chapter, suggested that SSA would be useless where critical thresholds are already formally adopted and followed as standards for sustainability goals, such as for goals established in national environmental policy plans. In such cases, it was indicated that no such tools, like SSA, are needed. In other words a sustainability trend is already in place and all decisions made at policy, planning, or project levels are surely sustainable.

If that is the case, then we will agree with that position, although it should be noted that SSAs main objective is not to define the sustainability thresholds, but rather to ensure, in a simple form, that the sustainability thresholds are effectively met in each incremental decision.

REFERENCES

1. European Commission, *Common Methodology for Multi-modal Trans-European Transport Networks - Deliverable 1*, COMMUTE-MEET Consortium, 1997.
2. European Commission, *European Sustainable Cities - Report by the Expert Group on the Urban Environment*, Ed. Office for Official Publications of the European Communities, Luxembourg, 1996.
3. Partidário, M.R., SEA - key issues emerging from recent practice, *EIA Rev.*, 16, 1996, 31–55.
4. European Commission, *Toward Sustainability - A European Community Programme of Policy and Action in Relation to the Environment and Sustainable Development*, II (5th EAP), 1993.
5. World Resources, *A Guide to the Global Environment*, World Resources Institute, Ed. Oxford University, New York, 1994–95.
6. United Nations Commission for Sustainable Development, *Indicators of Sustainable Development - Framework and Methodologies*, NY, 1996.
7. United Nations Conference on Environment and Development, EarthSummit '92, Rio de Janeiro - Agenda 21, Ed. Joyce Quarrie, London, 1992.

8. Commission of the European Communities, *The Future Development of the Common Transport Policy; A Global Approach to the Construction of a Community Framework for Sustainable Mobility*, COM (92) 494 Final, Brussels, 1992.

9. European Environment Agency, *Europe's Environment - the Dobris Assessment*, Eds. D. Stanners and P. Bourdeau. Office for Official Publications of the European Communities, Luxembourg, 1996.

10. UK Round Table on Sustainable Development, *Getting the Best Out of Indicators*, London, UK, 1997.

11. European Commission, *Strategic Policy Issues and Methodological Framework - Report by POSSUM project (Strategic Research Task 13)*, Policy Scenarios for Sustainable Mobility - Background Document of Deliverable 1, 1997.

12. George, C., Testing for sustainable development through environmental assessment, *EIA Review*, 19, 1999, 175–200.

APPENDIX

Selection of Indicators of Sustainable Development and Targets (Adapted from UNCSD (1996))

Category	Issue	Indicator		Target
SOCIAL	Promoting sustainable human settlement development	Driving force	Rate of growth of urban population	Not available
			Per capita consumption of fossil fuel by motor vehicle transport	Not available
			Human and economic loss due to natural disasters	Not available
		State	Percent of population in urban areas	Not available
			Area and population of urban formal and informal settlements	Not available
			Floor area per person	Not available
			House price to income ratio	Not available
		Response	Infrastructure expenditure per capita	Not available
ECONOMIC	International cooperation to accelerate sustainable development in countries and related domestic policies	Driving force	GDP per capita	Country specific
			Net investment share in GDP	Country specific
			Sum of exports and imports as a percent of GDP	Not available
		State	Environmentally adjusted Net Domestic Product	Not available
			Share of manufactured goods in total merchandise exports	Not available
		Response	None	

Selection of Indicators of Sustainable Development, and Targets (Adapted from UNCSD (1996)) (Continued)

Category	Issue	Indicator		Target
ENVIRONMENTAL	Conservation of biological diversity	Driving force	None	
		State	Threatened species as a percent of total native species	Threatened species are less than 1% of the total species in any class
		Response	Protected area as a percent of total area	10% of protected area for each major ecological region for countries by the year 2000.
INSTITUTIONAL	Science for sustainable development	Driving force	None	
		State	Potential scientists and engineers per million population	Not available
			Scientists and engineers engaged in R & D per million population	e.g., in Africa, 1 researcher per 1000 inhabitants by the year 2000.
			Expenditure on R & D as a percent of GDP	e.g., in Africa: - 1% of GDP by the year 1995 - at least 0,4%- 0,5% of GDP for research by the year 2000
		Response	None	

Section III

Procedural Approaches to
SEA at a Policy Level

4 Toward a Legal Framework for SEA in Canada*

Stephen Hazell and Hugh Benevides

CONTENTS

* This chapter is adapted from Hazell, S. and H. Benevides, Federal strategic environmental assessment: towards a legal framework, *J. of Environm. Law and Prac.*, 5, 1998.

LAW AND STRATEGIC ENVIRONMENTAL ASSESSMENT: INTRODUCTION TO THE ISSUES

The assessment of the environmental effects of proposed policies, plans, and programs (Strategic Environmental Assessment, or SEA) has potential as an approach to achieve sustainable development. The ultimate objective of SEA is to systematically integrate environmental considerations into government planning and decision-making processes relating to proposed policies, plans, and programs.

SEAs are rarely required as a matter of law in Canada or in other jurisdictions. Program EAs are required under the U.S. *National Environmental Policy Act*, and a directive on SEA is under consideration by the European Union.

The first issue to be explored in this chapter is whether or not a legal framework would promote SEA either by stimulating more widespread use of SEAs or higher-quality SEAs. A second issue concerns the possible elements of such a legal framework, assuming its desirability. The chapter explores these issues from a Canadian federal perspective by comparing the effectiveness of the 1990 directive of the Federal Cabinet, as described in the February 1993 document, *The Environmental Assessment Process for Policy and Program Proposals*, (the Cabinet Directive)[1] and the 1991 *Farm Income Protection Act* (FIPA)[2] in promoting SEA prior to 1998. Restated, the issue is whether either of these approaches work well in integrating environmental considerations into government planning and decision-making processes.

The Cabinet Directive requires, as a matter of policy, that federal departments assess the environmental effects of proposals for policies and programs requiring Cabinet approval. On the other hand, FIPA requires, as a matter of law, that environmental assessments be carried out for programs under FIPA agreements that provide financial support to Canadian farmers.

The chapter compares the elements of the Cabinet Directive and FIPA, the level of compliance by federal departments with these two sets of rules, and the quality of the SEAs carried out under each set of rules.

The chapter then discusses possible elements of a broader legal framework for federal SEA in Canada, reviewing key elements such as a basic legal requirement by federal departments to carry out SEAs, and provisions for public access to information about SEAs and reporting.

COMPARING THE CABINET DIRECTIVE AND THE *FARM INCOME PROTECTION ACT*: DOES EITHER APPROACH WORK WELL?

EA of proposed policies and programs of the Canadian federal government is not subject to any legally binding requirements, with the exception of those subject to FIPA. However, the Cabinet Directive does set out rules and guidelines for the conduct of SEAs for initiatives requiring approval of the Governor in Council (i.e., Federal Cabinet). The provisions of the Cabinet Directive and FIPA are compared below.

THE ENVIRONMENTAL ASSESSMENT PROCESS FOR POLICY AND PROGRAM PROPOSALS (STRATEGIC ENVIRONMENTAL ASSESSMENT CABINET DIRECTIVE)

The Cabinet Directive requires the EA of all federal policy and program initiatives submitted for Cabinet consideration.[3] Responsibilities of government ministers and officials in implementing the Cabinet Directive are detailed, and requirements for documentation, public statements, and public consultation are set out.

The objective stated in the Cabinet Directive is to systematically integrate environmental considerations into the planning and decision-making process. The environmental information derived from an examination of proposed policy or program initiatives is intended to support decision-making in the same way that other factors (i.e., economic, social, cultural) are now considered in evaluating proposals.

Policy and program proposals need not be assessed in emergency situations, where time is insufficient to undertake an EA, or when the Governor in Council is of the opinion that an EA would be inappropriate for reasons of national security.[4]

For environmentally relevant initiatives being considered by Cabinet, the Cabinet Directive provides that:

- a statement on environmental implications should be included in memoranda to Cabinet an,d where appropriate, in other documents submitted for consideration by ministers
- where anticipated environmental effects are likely to be significant, a more detailed account of the EA and the rationale for the conclusions and recommendations should be included in the documents supporting the proposal[5]

Under the Cabinet Directive, any disclosure of information is subject to existing legislation, regulations, and policies governing the release of information.

The Cabinet Directive also provides that ministers are to determine the content and extent of any public statement relating to the assessment according to the public interest and the particular circumstances of each case, where a statement is required.

The purpose of the public statement is to demonstrate that environmental factors have been integrated into the decision-making process, not necessarily to provide a detailed account of the assessment work undertaken.

The public statement need not always take the form of a separate document, but may be part of the announcement of the initiative or decision. The Cabinet Directive suggests that for initiatives likely to have significant environmental effects, public announcements should contain:

- a summary of the anticipated beneficial and/or adverse environmental effects of the initiative and their expected significance
- where relevent, information on measures adopted to mitigate adverse environmental effects, and on follow-up programs to monitor the initiative's effects over the longer term.[6]

The Cabinet Directive encourages public consultation for SEAs, but indicates that the nature and extent of public consultation is a matter of ministerial discretion.

FARM INCOME PROTECTION ACT (FIPA)

The 1991 FIPA requires an EA of programs established under the Act. The objective of FIPA is to authorize agreements between the Government of Canada and the provinces to provide income protection for producers of agricultural products. Section 4 of FIPA allows Cabinet to authorize the Minister to enter into an agreement with one or more provinces to provide for one or more farm income protection programs such as crop insurance.[7]

Section 4.(2) states that any program established under FIPA should encourage long-term environmental and economic stability. Section 5.(2) sets out EA and protection requirements for agreements respecting such programs, as follows:

An agreement respecting any program shall be subject to any applicable laws of Canada or a province,

a) providing for the circumstances and conditions under which insurance may be withheld, restricted, or enhanced for the purpose of protecting the environment and of encouraging sound management practices to ensure environmental sustainability; and

b) require an environmental assessment of the program be conducted within two years after the coming into force of the agreement and every five years thereafter, and provide for the manner in which the assessment is to be conducted

Thus, the agreements referred to Section 5.(2) providing for funding include provisions requiring the SEA to be conducted, and provisions setting out the manner in which the assessment is to be completed. This allows for a flexible approach to the SEA, because each agreement can be tailored to meet specific requirements. FIPA itself merely requires that an assessment be conducted within a legislated time frame.

Assessments are required to be conducted within two years of the agreement coming into force , and not prior to approval of the agreement, as might be expected. This approach contrasts with the Cabinet Directive, which is more in step with the usual EA principle, that EAs be carried out as early as is practicable in the planning stages, and before irrevocable decisions are made.[8]

FIPA provisions apply to a more limited range of government programs — agricultural subsidy under FIPA — than are covered under the Cabinet Directive, which is of a more general application. Further, the FIPA provisions do not touch on issues such as public consultation, the issuance of public statements summarizing the completed SEAs, and documentation and disclosure.

SEAs of the following programs have been carrried out by the Federal Department of Agriculture and Agri-Food under the provisions of FIPA:

- Gross Revenue Insurance Program (GRIP)
- Federal-Provincial Crop Insurance Program
- Net Income Stabilization Account (NISA)

In addition, two similar SEAs have been carried out by the Department:

- *Western Grain Transportation Act* (WGTA) Amendments
- Branch Rail Line Abandonment in Western Canada

Several other SEAs are in different stages of development.[9]

COMPLIANCE WITH THE CABINET DIRECTIVE AND THE *FARM INCOME PROTECTION ACT*

COMPLIANCE WITH THE CABINET DIRECTIVE

Estimating the level of compliance by federal departments with the Cabinet Directive is difficult given that little information about SEAs has been made available to the public. As indicated, the Cabinet Directive deals with submissions by ministers to the Cabinet. All such submissions, discussions, and documents relating to these submissions are subject to the rules of Cabinet confidence, which in turn is linked to the concept of Cabinet solidarity. This concept is that if Cabinet ministers are to accept a single government position on any given matter, they must have full confidence that they may express their views candidly in meetings without fear that their personal views may be publicly revealed to differ from the ultimate Cabinet decision. Such secrecy has hindered implementation of strategic EAs under the Cabinet Directive, because so little information about implementation is made public, and because there is no central point of access to the relevant information.

However, several sources of information can be brought to bear on the issue of compliance. These include:

- A January 1996 Canadian Environmental Assessment Agency[10] (CEAA) study examining the practices of federal departments in meeting the requirements of the Cabinet Directive
- The 1998 Report of the Commissioner of the Environment and Sustainable Development to the House of Commons[11]
- Various policy documents and guides that have been developed by federal departments in recent years demonstrating a commitment to SEA

The Agency study involved consultation with 20 federal departments and agencies on their implementation of the Cabinet Directive. The report found that some departments were applying SEA, conducting reviews, consulting the public, and issuing reports. Sixty percent of departments and agencies indicated that they had conducted SEAs, and 20% indicated that they had frameworks for accountability. Other departments:

- had "limited awareness of requirements"
- undertook " insufficient analysis and documentation" of SEAs
- "seldom" employed "environmental expertise and review" as part of the preparation of memoranda to Cabinet.[12]

The study further reported that departments found the Cabinet Directive to be "inadequate and confusing" and that "commitment at top level" was "lacking."[13]

This study indicates a low level of compliance with the Cabinet Directive, especially when one takes into consideration the natural tendency of officials to focus on positive aspects of departmental performance in such reviews.

Morever, all but a small handful of the SEAs carried out under the Cabinet Directive since June 1990 remain confidential. This is also suggestive of a low compliance rate, at least with respect to those provisions of the Cabinet Directive dealing with public reporting and disclosure. Of that small handful, to the knowledge of the authors, only two SEAs — the environmental reviews of the North American Free Trade Agreement (NAFTA)[14] and the Uruguay Round[15] — have been released publicly. Several other SEAs have been made available to the authors following their requests; still others have apparently been completed but remain confidential.

Given the government's prediction that perhaps as much as 25% of Cabinet's business is environmentally relevant and thus could require an EA under the Cabinet Directive,[16] a reasonable expectation would be that dozens, if not hundreds, of SEAs would be conducted each year. This has clearly not been the case.

The Commissioner of the Environment and Sustainable Development conducted an audit of the implementation of the Cabinet Directive, published in Chapter 6 of the May 1998 Report.[17] The Commissioner supported the findings of the 1996 CEAA study, concluding that "Departments have been slow to implement environmental assessment programs and policies."[18] Interviews conducted by the Commissioner's office found that:

- "most departments had not developed guidelines or directives"[19] on SEA

- "officials preparing these environmental assessments do not necessarily consult other departments with environmental expertise of their own experts in project environmental assessment."[20]

Interestingly, the Commissioner reports that: "departments may need additional pressure from parliamentarians, the public and the Commissioner of the Enviroment to fully implement the directive."[21]

There is evidence that in recent years federal departments are taking their responsibilities under the Cabinet Directive more seriously. This growing awareness may be a result of amendments to the *Auditor General Act,* which requires that departments prepare sustainable development strategies. The growing awareness may also be a result of the government's decision in September 1994 to strengthen the Cabinet Directive through non-legislative means.

The Agency has taken on a more vigorous role in advising federal departments on their responsibilities under the Cabinet Directive. A senior policy advisor at the Agency now has explicit responsibility for reviewing Cabinet submissions for compliance with the Cabinet Directive.[22] Further, the Agency, together with an interdepartmental working group, has prepared a draft, *Guide for Policy and Program Officers*, to provide practical advice for meeting the requirements of the Cabinet Directive.[23] In addition, the Agency and the working group have prepared an SEA training module based on the Guide to assist departments in training their policy and program staff. The Interdepartmental Committee on Policy and Program Environmental Assessment has also prepared recommendations for strengthening the SEA process at the federal level.[24]

The Federal Department of Environment has taken a leading role in terms of building its capacity to carry out SEAs and to provide scientific and technical advice to other departments that are preparing SEAs of their initiatives. The Department has sponsored several workshops on implementing SEA, and has developed a management framework on the division of roles and responsibilities within the department concerning SEA.

In addition to carrying out SEAs required under FIPA, the Department of Agriculture and Agri-Food's action plan under its sustainable development strategy commits the Department to:

1) review federal policies, programs, and initiatives aimed at rural Canada to ensure that agri-environmental interests are taken into account
2) enhance the capacity to conduct environmental analysis of new and existing policies and programs, and carry this out by:
    - providing training and guidance, including methods to improve consideration of biodiversity in conducting such analysis
    - conducting EA for policies and programs meeting the established criteria.[25]

In 1995, the Deputy Minister of Transport issued a circular entitled *Environmental Assessment (EA) of Proposed Policy and Program Initiatives in Transport Canada*[26] in response to the Cabinet Directive. The circular describes Transport

Canada's policy concerning SEA, and sets out a guide for practitioners in carrying out an SEA.

The Canadian International Development Agency (CIDA) has also produced a detailed *Guide to Integrating Environmental Considerations*[27] to help staff clearly and objectively integrate environmental considerations into their policy and program initiatives. The Department of Foreign Affairs and International Trade is preparing a similar guide, now in draft form.[28]

In summary, compliance with the Cabinet Directive to date by federal departments appears to be inconsistent at best, and very little information about specific SEAs has been released publicly; however, there is evidence that awareness and commitment of departments to SEA is slowly growing.

COMPLIANCE WITH THE *FARM INCOME PROTECTION ACT*

Compliance with the SEA provisions of the FIPA by Agriculture and Agri-Food Canada (AAFC) appears to be high, with five comprehensive SEAs completed over the past five years. The Auditor-General of Canada has recently reported that one federal-provincial program subject to FIPA was not subjected to an SEA; this SEA is currently being undertaken by AAFC.[29] Several other SEAs are currently being prepared. Although there is no legal requirement to do so, all of these SEAs have been tabled with Parliament and thus are available to members of the public.

COMPARISON OF THE QUALITY OF SEAS CONDUCTED UNDER THE CABINET DIRECTIVE AND THE *FARM INCOME PROTECTION ACT*

INTRODUCTION

In this section, the quality of SEAs carried out pursuant to the Cabinet Directive and FIPA are compared. This comparison of SEAs under the two sets of rules is exploratory and indicative only. The lack of publicly accessible SEAs under the Cabinet Directive in particular makes any definitive conclusions problematic.

The criteria chosen to compare the SEAs under the two sets of rules are as follows:

- *Scope.* Has the assessment been appropriately scoped? What baseline conditions have been adopted? Have important environmental issues been omitted?
- *Analysis.* Has an analysis been undertaken to examine environmental effects, test hypotheses, and generate predictions? What methodologies have been employed?
- *Alternatives.* Have alternatives to the proposed policy, plan, or program been elaborated and compared?
- *Mitigation and Follow-up.* Have mitigation and follow-up measures been considered and proposed?

- *Consultation.* Have the public or stakeholders been consulted in the preparation of the SEA?
- *Public Access to Information.* Have the SEA report and related documentation been made accessible to the public?

THREE STRATEGIC ENVIRONMENTAL ASSESSMENTS UNDER THE CABINET DIRECTIVE

Three SEAs under each set of rules were chosen to obtain a wide range of subject matters, and in the case of the Cabinet Directive, three different federal departments. The SEAs chosen under the Cabinet Directive are:

- the Environmental Review of NAFTA
- the new Minerals and Metals Policy (MMP) of Natural Resources Canada[30]
- a scoping study of the North American Waterfowl Management Plan (NAWMP)[31]

The North American Free Trade Agreement

The *North American Free Trade Agreement: Canadian Environmental Review* (NAFTA SEA) was published in October 1992, less than two months before NAFTA was signed.

According to the NAFTA SEA, the environmental review had two objectives: "to improve decision making by identifying opportunities to maximize environmental benefits and to avoid or minimize negative impacts" and to "provide information on the environmental effects and related consequences of alternatives, so that environmentally responsible choices can be made from among the various options available."[32] The objectives included in the terms of reference of the environmental review committee were narrower: "to ensure that environmental considerations were taken into account during all stages of the negotiating process; and to conduct and document a review of the potential environmental effects of the NAFTA on Canada."[33]

The environmental review committee included representatives of federal departments, such as: Foreign Affairs and International Trade; Agriculture; Energy, Mines and Resources; Environment; Finance; Fisheries and Oceans; Forestry; Industry, Science and Technology; and Transport. A "technical expert advisory group" from the Department of Environment provided support.

The committee was charged with assembling and reviewing relevant literature, meeting with Canadian negotiating team members to discuss the scope of the negotiations in each group, and consulting with advisory groups, provincial officials, and environmental reviewers in the U.S.and Mexico.

The environmental review was required to concentrate on potential environmental effects of NAFTA on Canada, "while responding to the full range of concerns expressed by Canadians."[34] It was to address effects as relevant to each negotiating group, on the basis of the ecological media (soil, air, and water), and to comment on activities or measures undertaken in parallel discussions with the other NAFTA members and in other relevant fora.[35] Consistent with the Cabinet Directive, the review was required to be submitted to the Cabinet no later than the date of the signing of NAFTA itself.

Minerals and Metals Policy

The *Environmental Assessment of the new Minerals and Metals Policy* (MMP SEA) was prepared by the MMP Branch, the Minerals and Metals Sector, of Natural Resources Canada in 1996. The purpose of the SEA was to "provide information on the potential impact of two options to address the environmental, social and economic aspects of federal policies, programs and legislation that relate to Canada's minerals and metals industry."[36] These two options were to: continue to operate under the 1987 MMP of the Government of Canada; or develop a new policy based on the concept of sustainable development. Environmental, social, and economic implications of the two options were reviewed using a methodology that involved a literature review and consultations with federal, provincial, industry, environmental, and aboriginal stakeholders.

North American Waterfowl Management Plan

The North American Waterfowl Management Plan (NAWMP) EA Scoping Study was carried out for the Environmental Assessment Branch of the Department of Environment in 1991. The scoping study considered the NAWMP, which was initiated in 1986 as a continent-wide, multi-year initiative of habitat protection, enhancement, and restoration measures to counter the decline in waterfowl populations observed over the previous quarter-century.

The study was limited to the scoping stage and did not include a final EA (or at least no final SEA is publicly available). The members of *The NAWMP EA Scoping Study* team included representatives of the Department of Environment, including administrators of NAWMP, the Canadian Wildlife Service, AAFC, the Federal Environmental Assessment Review Office (FEARO), and the consulting team.

THREE STRATEGIC ENVIRONMENTAL ASSESSMENTS UNDER THE *FARM INCOME PROTECTION ACT*

The three SEAs chosen under FIPA are:

- Net Income Stabilization Account (NISA)[37]
- Federal-Provincial Crop Insurance Program[38]
- WGTA Amendments[39]

Net Income Stabilization Account (NISA)

An SEA of NISA was conducted for AAFC in September 1993 in accordance with the requirements of the FIPA and the Federal-Provincial Crop Insurance Program. NISA is a self-insurance fund for Canadian farmers. Contributions are made by the federal and provincial governments as well as the producers themselves, allowing producers to stabilize their income during low-income years. Contributions are limited to eligible sales of grains, oilseeds, most specialty crops, and edible horticultural crops. The objective of the SEA was to determine whether NISA, during its first two years of operation, had influenced farmers to modify their crop selection, land use, chemical use, and management practices in a way that would harmfully impact wetlands, wildlife habitat, soils, water quality, or air quality. The SEA report was tabled in Parliament in 1994.

Federal-Provincial Crop Insurance Program

Some form of crop insurance has existed in Canada for more than 20 years, to protect farmers against yield losses caused by natural hazards. Funded by payments from producers and the federal and provincial governments, and administered jointly by both levels of government, crop insurance guarantees payment to farmers based on a minimum yield. The Federal-Provincial Crop Insurance Program was implemented under the FIPA in 1991, and the federal-provincial agreements that followed.

The objective of the SEA of the Federal-Provincial Crop Insurance Program was to determine the environmental effects attributable to the existence, structure, and delivery of crop insurance in Canada. An SEA of crop insurance was completed for AAFC in March 1994. The report was distributed to federal and provincial bodies responsible for the administration and reform of crop insurance programs.[40]

Western Grain Transportation Act (WGTA) Amendments

Statutory freight rates for the shipment of grains, first introduced in 1897, and various freight subsidies, in place since the 1980s, have been an integral part of agricultural policy in Canada. Subsidies under the WGTA, first enacted in 1983, replaced the so-called "Crow Rate" for shipping prairie grain. The transportation subsidy was terminated in August 1995. The method of payment of federal subsidies was a much-debated issue, both because various options would have favored certain agricultural producers over others, and because they would have had direct impacts on farm practices.

In 1992, an assessment of the environmental impacts of proposed amendments to the WGTA was prepared for AAFC. The SEA considered the potential for land-use changes and associated impacts on the air, water, and terrestrial ecosystems relating to three alternative-proposed grain transportation policies that were under review at that time. The environmental effects associated with increased trucking of grain resulting from rationalization of the grain transportation system were assessed in a separate study.

SCOPE

This section examines the appropriateness of the scoping undertaken, and baselines established under each of the six SEAs.

Cabinet Directive

NAFTA SEA: The NAFTA SEA purported to consider four categories of issues:

- The environmental relevance of selected provisions of the Agreement
- The potential impact of NAFTA on Canada's environment
- The prospects of "industry migration" from Canada to places with less stringent environmental regulations
- Follow-up mechanisms intended to allow trade and environment issues to be considered following signature of the Agreement

The SEA provided little information as to whether all environmental issues related to NAFTA were included in the review, or how the issues actually included were selected. For example, the section examining the issue of "industry migration" referred to approximately a dozen studies or reports on aspects of the issue. A literature review of trade and environment issues, even restricted to 1992 and earlier, would probably reveal a much larger body of literature. The SEA provided no guidance on how the issues discussed in the review were selected.

The NAFTA SEA included several annexes describing the Canada-Mexico and U.S.-Mexico trade climate in support of a baseline for assessment of "freer trade" between Canada and the U.S. This baseline was justified on the basis that the Canada-U.S. Free Trade Agreement had been in effect for some years at that time.

MMP SEA: The existing MMP was adopted as the baseline for the new policy. The SEA did not attempt to explicitly scope the environmental issues to be analyzed in the SEA, nor were industrial activities or geographic applications scoped.

NAWMP EA Scoping Study: Part of the scoping process was an elaboration of the characteristics of the policy/program (the NAWMP initiative was considered by the reviewers to be part policy, part program). For example, the NAWMP scoping study attempted to characterize NAWMP. Some of the characteristics mentioned were that it covered a large geographic area, and that it was implemented by regional joint-ventures, including broad governmental and non-governmental organization representation, each administered somewhat differently.

Farm Income Protection Act

NISA SEA: The SEA was scoped based on a detailed analysis of relevant environmental issues at the national and regional levels in Canada. A series of questions were then developed to focus the analysis. The baseline for the analysis was pre-NISA environmental effects, which were then compared to post-NISA effects. The SEA states that it would have been unrealistic to generate a baseline scenario founded on assumptions about what environmental effects might have occurred in the absence of any farm income protection programs.

Crop Insurance SEA: The baseline employed for this SEA was a status quo condition for all other agricultural programs in the absence of crop insurance. It was argued that environmental effects associated with the Federal-Provincial Crop Insurance Program could thereby be isolated more easily. As part of the scoping of the review, the project team, in conjunction with a steering committee and other technical experts, identified key environmental issues such as how crop insurance would affect soil quality, wildlife habitat, and production decisions generally.

WGTA SEA: The SEA was scoped by reference to the geographic region of impact and the environmental factors considered. The geographic scope of the study was the agricultural region of the prairie provinces, focusing on those areas where the greatest land-use changes were predicted. The key environmental factors identified by a scoping analysis were soil degradation, surface and ground water quantity, wildlife and wildlife habitat, and air and climate. Socio-economic impacts were not assessed.

ANALYSIS

This section briefly reviews the methodologies employed to examine environmental effects, whether or not any hypotheses were tested as part of the analysis, and whether or not predictions relating to environmental effects were generated through the use of modeling or other means.

Cabinet Directive

NAFTA SEA: The methodology employed in the NAFTA SEA was essentially a literature review combined with limited consultations with experts and stakeholders. There does not appear to be any testing of hypotheses, except to the extent that some literature in which hypotheses were tested was reviewed, nor were any predictions offered about environmental effects resulting from NAFTA. The analysis of environmental effects was limited, possibly due to a lack of information about the relationships between trade and environment.

MMP SEA: Analysis of possible environmental effects resulting from the proposed MMP compared to the existing policy was limited. The MMP SEA argued that the existing MMP did not reflect the current policy environment in that it did not integrate sustainable development as that concept was coming to be understood. However, the SEA did not elaborate on the details of the proposed policy directions and the environmental effects these proposed policy directions might have had. The assumption appeared to be that the new policy would be a cure-all because of its sustainable development orientation. This sort of analysis begs the question of a strategic EA: "what is the best option?" Or even, "what are the relative merits of this policy option?"

While the MMP SEA listed "References" in Appendix A, it did not indicate the sources of the ideas contained in the text of the SEA. Thus, sentences like "... the challenge of sustainable development in the context of minerals and metals is considered as incorporating the following elements . . .".[41] do not reveal the authority for the "considered" proposition.

NAWMP EA Scoping Study: As already noted, the NAWMP report reviewed was a scoping study; as such, it included an extensive listing and description of environmental effects (using a broad definition of "environment"). It also included a thorough discussion of process and design issues such as consultation requirements, the problems posed by the unique organizational structure of NAWMP, as well as resource constraints. It was beyond the scope of the study to analyse environmental effects associated with different policy options.

Farm Income Protection Act

NISA SEA: The analysis for this SEA relied on several different methodologies including:

- theoretical and conceptual assessment (employing economic analysis of agricultural subsidies generally, and analysis of likely effects of NISA on production choices)
- descriptive assessment (an analysis of comments received from NISA participants, as well as other interested parties)
- simulation studies to assess soil erosion
- synthesis, conclusions, and recommendations

The SEA made specific determinations about the nature of environmental effects (e.g., positive/negative; short-term/long-term; negligible/major) for different environmental factors (e.g., soil erosion, water quality) related to NISA.

Crop Insurance SEA: The analysis in the SEA comprised four stages or components, namely:

- literature review and analysis
- aggregate analysis of data relating to participation, acreage, and input use
- farm-level decision models using hypothetical producers under a range of conditions
- qualitative analysis of feedback solicited from participating farmers and of environmental effects of structural differences among provincial crop insurance programs.

Environmental impacts were assessed for each of the nine environmental issues identified at the scoping stage of the analysis. As for the NISA SEA, specific judgements were made concerning the nature and significance of these environmental impacts. In addition, the study analyzed how crop insurance fit within the broader context of government programs and market signals in influencing decisions of agricultural producers.

WGTA SEA: The analysis involved a methodology that included both a literature review as well as quantitative and qualitative assessments, in line with the limitation on the availability and quality of existing data and information used.

The Canadian Regional Agricultural Model (CRAM) was employed as a tool for various enterprises in three broad categories of economic change (i.e., change in enterprise mix, change in cropping patterns and intensity, change in input levels). The model provided quantitative assessments on the magnitude of land-use change, the theoretical risk of soils, water, wildlife, and air to each land use type, and the sensitivity of each crop district to change.

Again, assessments of the nature and significance of environmental impacts were provided for each of the four options under the WGTA.

ALTERNATIVES

Cabinet Directive

The Cabinet Directive does not specifically require consideration of policy or program alternatives, nor of the anticipated environmental aspects of these alternatives.

> *NAFTA SEA:* This SEA included no information suggesting that policy alternatives to NAFTA were considered. The terms of reference for the SEA specifically provided that the purpose of the SEA was not to re-examine trade objectives. However, the SEA did state that one of the principal objectives of environmental review was to "provide information on the environmental effects and related consequences of alternatives."[42]
>
> *MMP SEA:* According to the MMP SEA, the only alternative policy to the new MMP that was considered was the status quo: the existing (1987) MMP of the Government of Canada. Alternatives to the proposed policy favored by the government based on the SEA or the consultations were not laid out in the SEA. It is difficult to determine from the documentation whether or not the SEA had any impact on the development of the new policy.
>
> *NAWMP EA Scoping Study:* As a scoping study, the NAWMP report did not consider alternatives.

Farm Income Protection Act

> *NISA SEA:* The SEA sets out a number of alternative approaches and options for measures in the delivery of NISA to achieve sustainable agricultural objectives. These alternatives flowed from the analysis described earlier.
>
> *Crop Insurance SEA:* The SEA of the Federal-Provincial Crop Insurance Program considered alternatives by discussing a variety of production

choices and their likely environmental effects using a range of analytical tools, as described above.

WGTA SEA: The report examined five policy alternatives for amending the WGTA. Three alternatives were studied using quantitative and qualitative approaches, while a fourth was studied on a qualitative basis only. An alternative, which assumed no agricultural production activity, was not considered since it would not have been feasible or realistic.

MITIGATION AND FOLLOW-UP

Cabinet Directive

NAFTA SEA: Mitigation was included in the NAFTA review only insofar as environmental issues were discussed, as noted above. A discussion on follow-up noted that "the NAFTA negotiations acted as a catalyst to increased environmental co-operation between Canada and Mexico under the parallel process" and to increased sophistication in discussions of trade-environment linkages. In keeping with its descriptive nature, the review also noted that provisions of the NAFTA would require "the continued consideration of the relationship between trade and the environment within the context of the Agreement."[43]

There were no follow-up measures included as part of the NAFTA SEA, although reference is made to the North American Agreement on Environmental Cooperation (NAAEC), a so-called side-agreement to NAFTA. NAAEC was signed by Canada, Mexico, and the U.S. as an approach to address some of the poorly understood links between trade and environment in conjunction with NAFTA.

MMP SEA: There were no mitigation or follow-up measures included in the MMP SEA.

NAWMP EA Scoping Study: The NAWMP Scoping Study pointed out the need for monitoring and evaluation of results of the SEA. As a scoping study, recommendations concerning mitigation or follow-up measures would not have been expected.

Farm Income Protection Act

NISA SEA: A number of recommendations relating to mitigation and follow-up were provided in the SEA together with recommendations for environmental monitoring programs dealing with land use, soil degradation, non-point source pollution, wildlife habitat changes and surface water quality. In addition, information deficiencies and future study needs were identified.

Crop Insurance SEA: The SEA set out a number of recommendations relating to mitigation and follow-up. The report included recommendations to define minimum guidelines for good farming practices for incorporation into provincial crop insurance programs, monitoring of environmental effects to be carried out by producers and program administrators, testing of modifications to crop insurance prior to full adoption into the program, and monitoring and development of inventories of marginal or environmentally sensitive lands and habitats, among others.

WGTA SEA: The SEA provided recommendations relating to environmental monitoring and data collection on farms as part of a follow-up program. The SEA also pointed out a number of limitations to the study.

CONSULTATIONS

Cabinet Directive

The level of consultations carried out for the three SEAs varied considerably.

NAFTA SEA: The consultative process for the NAFTA SEA was conducted through the International Trade Advisory Committee (ITAC) and several Sectoral Advisory Groups on International Trade (SAGITs), which comprised members of business, environment, labor, and academic sectors, and was reported directly to the Trade Minister. These ITAC and SAGIT chairs and environmental representatives met with the senior negotiation team for one day in February 1992 to discuss the environmental implications and the content of the review. The same committee members, along with some other environmental groups, were invited to a further workshop on trade and environmental issues, held in April 1992.

One outcome of the April meeting was a recommendation, subsequently met, that the terms of reference of the environmental review should be made available to the public. At a third meeting in September 1992, the environmental provisions of NAFTA and the structure of the review were presented to ITAC and SAGIT members plus some environmental groups. Several submissions of Canadian and other environmental groups were also considered by the review committee for discussion with the negotiators. Provincial officials were consulted through a Federal-Provincial Committee on NAFTA.[44]

MMP SEA: Natural Resources Canada produced an issues paper, which formed the basis for consultations on the MMP. Industry, environmental, and aboriginal groups, federal and territorial governments, and parliamentarians, were consulted based on this issues paper. However, to all appearances, the SEA itself was drafted and reviewed internally, with no consultation or review by members of the public or stakeholders.

NAWMP Scoping Study: By admission of the NAWMP scoping study report itself, "consultations were not exhaustive" and were not extensive. An appendix listed the persons consulted; these were NAWMP officials, representatives of federal departments, and so-called "program EA specialists."[45] Apparently, no public consultations were conducted. The NAWMP Scoping Study included a short discussion of the need for consultation with potentially affected groups and organizations as well as potential partners, and pointed out the need for an eventual program assessment to provide a forum for such stakeholders to participate and allow NAWMP to accomplish improved environmental performance.

Farm Income Protection Act

NISA SEA: A mail survey was distributed to 10,000 farmers across Canada to identify NISA-related environmental issues. In addition, structured interviews were held with resource professionals across Canada to solicit opinions on NISA, and further contributions were obtained through written submissions from the public and interested organizations.

Crop Insurance SEA: Extensive consultations were held with a broad range of stakeholders to garner their perspectives on both the importance of environmental issues in agriculture and the role that crop insurance plays in affecting these issues. A request from the Federal Minister of Agriculture was widely distributed and additional in-person interviews were conducted with many of the stakeholders.

WGTA SEA: The SEA does not indicate that consultations were held either with stakeholders or the public. The SEA report does recommend a "bottom-up planning process which recognizes the knowledge of individual farmers as the primary database to drive land allocation and management processes..."[46]

PUBLIC ACCESS TO INFORMATION

Cabinet Directive

NAFTA SEA: The NAFTA SEA was published in October 1992 and is publicly available through the Canada Communications Group.

MMP SEA: This SEA was made available by Natural Resources Canada to the authors at their request on the condition that it not be reproduced as an annex to this chapter. Otherwise, the SEA is not publicly accessible. Apparently, no public statement as called for under the Cabinet Directive was released following the government's decision to approve the policy.

NAWMP EA Scoping Study: This study was made available by Environment Canada officials to the authors at their request. The SEA is available to the public, but is not easily accessible. Apparently, no public statement as called for under the Cabinet Directive was released following the government's decision to approve the renewal of NAWMP.

Farm Income Protection Act

All three SEAs are publicly available as they have been tabled with Parliament. However, these SEAs are not easily accessible by members of the public. Indeed, few federal officials involved in SEA outside of AAFC appear to be aware of the FIPA or the SEAs conducted pursuant to its provisions.

CONCLUSIONS

From the above comparison of the Cabinet Directive and FIPA, several conclusions can be drawn regarding whether a legal framework works well in supporting SEA, or whether a policy framework alone is preferable.

To review, FIPA requires SEAs as a matter of law and the Cabinet Directive, as a matter of government policy. The FIPA SEA provisions are more narrowly drafted and do not provide for or require public consultation or reporting, although SEAs under FIPA are typically tabled in Parliament and thus are publicly available. SEAs under FIPA are usually conducted after the farm income protection program has been implemented, whereas the Cabinet Directive requires that SEAs be conducted before decisions are made.

Compliance with FIPA since 1991 has been high, whereas compliance with the Cabinet Directive since 1990 has been inconsistent at best. Two reasons are suggested for this difference. The first more obvious reason is that FIPA SEAs are required by law, whereas Cabinet Directive SEAs are not. The second reason relates to the timing of the SEA. Carrying out the SEAs after the government has made its decision means that AAFC does not need to be concerned about the Cabinet confidentiality considerations that limit SEAs carried out as part of the Cabinet decision-making process.

Second, the three SEAs conducted under FIPA were generally superior in quality to the three conducted under the Cabinet Directive, based on the criteria described in Section 4. The scoping of environmental issues and other elements of the SEAs under FIPA was more systematic and thorough, and the methodologies employed in analyzing possible environmental effects were superior, especially in their deployment of sophisticated models and use of multiple approaches. The elaboration of policy or program alternatives was better developed than for the Cabinet Directive SEAs, and mitigation and follow-up measures were better considered and laid out. With respect to consultation, neither set of SEAs was remarkably superior to the other. Public access to information about FIPA SEAs was generally better than for those under the Cabinet Directive. But in neither case were the SEAs easily accessible to the public (with the exception of the NAFTA SEA).

The reasons for the superiority of FIPA SEAs may be the same as those suggested above with respect to compliance. It may at first seem counterintuitive for a law with few requirements to produce better results than a non-legally binding policy with additional requirements. However, the impact of a law on government officials may inspire more respect, in part because of greater public accountability and transparency, and thus the desire to see the work done well. The after-the-fact nature of FIPA SEAs takes the specific SEA out of the policy and decision-making loop for that program, and thus allows greater involvement of stakeholders and public reporting of the SEAs.

· The after-the-fact nature of SEAs under FIPA runs contrary to a key principle of EA — that EAs be undertaken as early in the planning process as possible and before irrevocable decisions are made. However, this principle was developed in the context of project EAs, which typically are discrete and "one-off" — the project is planned, initiated, completed, and operated. Policies, plans, and programs differ from projects in that they are rarely new and rarely end. However, new proposals for policies, plans, and programs are incremental, building on what has gone before, through various processes of reforming, adding, and subtracting different elements. The federal government has had trade and mining policies, and wildlife and farm income protection programs in a myriad of forms for decades.

Given that the making of policies, programs, and plans is essentially incremental in its nature, the requirement that SEAs be carried out before decisions are made becomes less essential, in part because the decisions are not irrevocable. Once a dam is built, the environmental harm cannot easily be undone, unless it is removed. Governments can and do change their policies, plans, or programs, easily in some cases (e.g., budgets change every year), and with greater difficulty in others (e.g., international trade agreements such as NAFTA).

For their part, farm income protection measures are constantly in flux in Canada. SEAs completed for one program initiative may not affect the decision-making for that initiative, but likely will affect succeeding initiatives. An interesting after-the-fact aspect of FIPA SEAs is that experience and expertise in undertaking SEA are allowed to build; they represent a commitment to sustainable agriculture over a longer term.

WAYS FORWARD

The albeit-limited comparisons undertaken in this chapter suggest that a legal framework may be desirable from the perspective of improving compliance with SEA rules across government departments, as well as improve the quality of SEAs conducted under those rules. Suggested principles for any federal SEA statute would be to have a minimum of legal requirements, afford a maximum of flexibility to federal departments to integrate the EA activity into their decision-making processes, and employ existing institutions (e.g., Canadian Environmental Assessment Agency, departmental environmental assessment teams) to minimize the costs of administering the new legislation. Assuming that a legal framework is desirable, the elements of the framework might include:

- a legal requirement for federal authorities to conduct SEAs of proposed policies, plans, or programs likely to have adverse environmental effects
- a provision providing federal authorities with the option of conducting the SEA following the decision of the Governor in Council to proceed with the proposal, where the federal authority determines that an after-the-fact SEA is in the public interest
- few provisions dealing with the factors to be considered or the elements of the SEA, allowing federal authorities the flexibility to adapt the SEA to the existing policy-making processes. Provisions could include requirements to summarize the anticipated beneficial and/or adverse environmental effects of the proposals and their expected significance, to provide information on measures adopted to mitigate adverse environmental

effects, and to monitor the proposal's effects over the longer term on follow-up programs

- a legal requirement that federal authorities provide a public registry maintained by the Canadian Environmental Assessment Agency with a summary of the SEA, at a minimum, where it has been conducted as part of the decision-making process; and the complete SEA where an after-the-fact SEA has been conducted
- provisions for the administration of the SEA regime

REFERENCES

1. Federal Environmental Assessment Review Office, *The Environmental Assessment Process for Policy and Program Proposals*, FEARO, Hull, Quebec, 1993.
2. S.C. 1991, Chapter 22.
3. Supra, Note 2, 5.
4. Supra, Note 2, 4.
5. Supra, Note 2, 6.
6. Supra, Note 2, 7.
7. Supra, Note 3, Section 4.
8. Section 5(2)(b)(i) of *The Canadian Environmental Assessment Act.*
9. Ian Campbell, AAFC, personal communication.
10. Canadian Environmental Assessment Agency, *Review of the Implementation of the Environmental Assessment Process for Policy and Program Proposals*, CEAA, January 1996.
11. Commissioner of the Environmental and Sustainable Development, *Report of the Commissioner of the Environment and Sustainable Development to the House of Commons*, Minister of Public Works and Government Services, Canada, 1998.
12. Commissioner of the Environmental and Sustainable Development, *Report of the Commissioner of the Environment and Sustainable Development to the House of Commons*, Minister of Public Works and Government Services, Canada, 1998.
13. Commissioner of the Environmental and Sustainable Development, *Report of the Commissioner of the Environment and Sustainable Development to the House of Commons*, Minister of Public Works and Government Services, Canada, 1998.
14. Government of Canada, North American Free Trade Agreement: Canadian Environment Review, *Environ.*, Canada, October 1992.
15. Government of Canada, *Uruguay Round of Multilateral Trade Negotiations: Canadian Environmental Review*, Department of Foreign affairs and International Trade, April 1994.
16. Supra, Note 2, 5.
17. Supra, Note 12.
18. Supra, Note 12, 6-26.
19. Supra, Note 12, 6-27.
20. Supra, Note 12, 6-27.
21. Supra, Note 12, 6-27.
22. Dr. Bill Couch, CEAA, personal communication.
23. Canadian Environmental Assessment Agency, *Guide for Policy and Program Officers*, CEAA, 1977.

24. Interdepartmental Committee on Policy and Program Environmental Assessment, *Strengthening Strategic Environmental Assessment*, CEAA, 1996.
25. Agriculture and Agri-Food Canada, *Agriculture in Harmony with Nature: Strategy for Environmentally Sustainable Agriculture and Agri-food Development in Canada*, Minister of Public Works and Government Services, Ottawa, Ontario, 1997, 9, 10.
26. Deputy Minister of Transport, *Environmental Assessment (EA) of Proposed Policy and Program Initiatives in Transport Canada*, Transport, Canada, 1995.
27. Canadian International Development Agency, *Guide to Integrating Environmental Considerations*, CIDA, 1996.
28. Shuttleworth, J. and J. Howell, Strategic Environmental Assessment within the Government of Canada and specifically within the Department of Foreign Affairs and International Trade (Paper presented to the 1997 Conference of the International Association of Impact Assessment, May 1997).
29. Ian Campbell, AAFC, personal communication.
30. Mineral and Metal Policy Branch, Minerals and Metals Sector, Natural Resources Canada, *Environmental Assessment of the New Minerals and Metals Policy* (undated).
31. Lavalin Environment (1991) and ARA Consulting Group, Inc., *NAWMP EA Scoping Study*, February 1992.
32. Supra, Note 15, 3-4.
33. Supra, Note 15, 3-4.
34. Supra, Note 15, 75-76.
35. Supra, Note 15, 76.
36. Supra, Note 31.1.
37. Environmental Management Associates, *Environmental Assessment of NISA: Final Report*, Calgary, Alberta, September 1993.
38. Price Waterhouse, *Synthesis and Recommendations: Environmental Assessment of Crop Insurance, Final Report*, March, 1994.
39. Terrestrial & Aquatic Environmental Managers Ltd., *An Environmental Assessment of Land Use Changes Due to Proposed Modifications to the Western Grain Transportation Act*, Melville, Saskatchewan, December 1992.
40. Campbell, I., SEA: A Case Study of Follow-Up to Canadian Crop Insurance in *The Practice of Strategic Environmental Assessment*, Therivel, R. and Partidario, M.R., Eds., Earthscan Publications, London, U.K., 1996.
41. Supra, Note 31, 9.
42. Supra, Note 17, 3.
43. Supra, Note 15, 73.
44. Supra, Note 15, 5-6.
45. Supra, Note 32, Appendix 2.
46. Supra, Note 40, 74.

5 SEA Within the Government of Canada and Specifically Within the Department of Foreign Affairs and International Trade

Jaye Shuttleworth and Jennifer Howell

CONTENTS

INTRODUCTION

This chapter outlines some of the tools that the Canadian Environmental Assessment Agency (Agency) and the Canadian Department of Foreign Affairs and International Trade (DFAIT) have developed to assist federal departments and DFAIT specifically to integrate environmental assessment (EA) considerations into policy and program development.

A manual entitled, *Strategic Environmental Assessment: A Guide for Federal Policy and Program Officers* (January 1997), was prepared by the Agency under the guidance of an interdepartmental working group on strategic environmental assessment (SEA). Concurrently, the Department of Foreign Affairs and International Trade developed specific tools and processes to assist policy officers in evaluating the environmental consequences of proposed initiatives. This paper will include an explanation of those tools, including:

- the generic *Strategic Environmental Assessment: A Guide for Federal Policy and Program Officers* (January 1997) and supporting training module
- the DFAIT *Environmental Implications Checklist* (addresses whether an assessment is required and the level of assessment required)
- the DFAIT *Guide to Completing Environmental Implications Checklist* (addresses some of the methodology and analysis required to complete an EA)

Most departments within the Canadian government are attempting to find ways to assess the environmental effects of proposed policies and programs. This chapter will describe the federal requirements for Strategic Environmental Assessment (SEA), some of the benefits (environmental and economic) of SEA and the DFAIT's initiatives to develop and promote a systematic SEA process.

A previous version of this paper was delivered at the annual conference of International Association of Impact Assessment in New Orleans in May 1997.

WHAT ARE THE CANADIAN REGULATORY REQUIREMENTS FOR STRATEGIC ENVIRONMENTAL ASSESSMENT?

SEA is evolving in many countries as an important decision-making tool for sustainable development. In Canada, the importance of SEA has become more evident with the creation of the Office of the Commissioner for Environment and Sustainable Development; the passing of amendments to the Canadian Environmental Assessment Act, which provides the legal framework for conducting EAs of projects involving a federal government decision; and recent initiatives led by the Agency to provide advice and guidance to departments on SEA.

As a result of the 1990 Cabinet Directive, federal departments and agencies are required to consider environmental factors in many government policy and program proposals.

In 1993, the Agency issued procedural guidance in a brief report entitled, *The Environmental Process for Policy and Program Proposals*. According to these instructions federal departments are required to:

- systematically integrate environmental considerations into the planning and decision-making process of policy and program proposals submitted for Cabinet consideration or considered by ministers or their own authority, at the discretion of the responsible minister
- outline, when appropriate, the environmental effects considered in Memoranda to Cabinet (MC) and relevant documents
- prepare, when appropriate, a public statement demonstrating that environmental factors have been integrated into the decision-making process
- consult, when appropriate, with stakeholders and the general public

Individual ministers are responsible for ensuring that their departments comply with the conditions of the 1990 Cabinet Directive. However, the process is based on the principle of self-assessment and ministers have the flexibility to apply approaches suited to their department's unique needs and circumstances.

Roughly two years later, a review by the Agency of the work of federal departments in implementing the process, indicated that the Directive was poorly understood and was being applied in an inconsistent manner. The review also indicated the need for further procedural guidance and examples.

In order to draw on the experience of federal departments, the Agency established an Interdepartmental Working Group on SEA. The working group was comprised of experienced EA practitioners, policy analysts, and program developers from 14 federal departments and agencies. It provided guidance and direction in the preparation of two documents: *Strategic Environmental Assessment: A Guide for Policy and Program Officers* and the *Environmental Assessment of Policies, Programmes and Plans: A Training Manual*. These documents offer a simple, practical approach that can be integrated into departmental policy/program development processes, and a generic SEA guide and training material can be used by policy and program officers responsible for conducting SEAs of proposed policies or programs.

The approach is one of "train the trainers," and involves the following five questions:

1) How can you integrate environmental considerations as early as possible in the policy/program process?
2) What are the possible environmental considerations of the policy/program options?
3) What are the likely stakeholder concerns?
4) How can you inform decision-makers?
5) What will you need in place after the decision?

The working group meetings were very much one of consensus-building. Members were asked to share their experiences and perspectives in a cooperative exercise aimed at providing practical, effective guidelines and advice that reflected the experiences of policy and program officers and EA practitioners. The guidelines also had to reflect the reality that without a legislative base requiring strategic EA, its development and application at the federal level could best be accomplished by 'persuasion,' and in the provision of useful examples.

The training module includes an overview of SEA, a complete presentation deck and speaking notes to assist the trainers, a dozen exercises with supporting information for trainers, and a reference section for those wishing additional material.

The module has been well received in many respects because of the need, arising from recent amendments to the Auditor General's Act (Bill C-58), for federal departments to develop sustainable development strategies with accompanying action plans for implementation and performance indicators so that the implementation of each strategy can be monitored. The Commissioner for Environment and Sustainable Development will be auditing each department's performance, and elected representatives at the federal level will be held responsible for the outcome of these audits.

The *Strategic Environmental Assessment: A Guide for Policy and Program Officers* and *Environmental Assessment of Policies, Programmes and Plans: A Training Manual* are extremely useful, but apply, in the most part, to proposals that are relatively modest and routine in nature. The *Guide* itself acknowledges that "one-time large scale, controversial policies with national implications such as international trade agreements, likely will require more thorough and specialised analysis of all factors, not only environmental concerns."

As these types of proposals fall within the mandate of the DFAIT, the Department determined that it would develop an internal process that would build on existing approval systems and focus on improving employee awareness.

THE DEPARTMENT OF FOREIGN AFFAIRS AND INTERNATIONAL TRADE'S COMMITMENT TO SUSTAINABLE DEVELOPMENT AND STRATEGIC ENVIRONMENTAL ASSESSMENT

As per amendments to the Auditor General's Act, the Department tabled *Agenda 2000*, its first sustainable development strategy, in Parliament in December 1997. The strategy committed the Department to conduct environmental reviews of all recommendations to Cabinet.

Given that the integration of environmental concerns into policy development within DFAIT is critical to the Department's sustainable development objectives, the Department realized that appropriate SEA tools had to be developed that recognized both the challenge of implementing SEA in the international area and the unique circumstances of the Department.

IMPLEMENTATION CHALLENGES

Some of the issues that had to be addressed when developing appropriate tools for use by the Department included the following:

THE ROLE OF THE DEPARTMENT

The Department represents the Government of Canada in international negotiations and agreements and has responsibility for signing international agreements on behalf of Canada. As such, many of the policy issues negotiated by DFAIT are the direct responsibility of other departments who are usually in charge of implementing the provisions of the agreements.

PREVIOUS ENVIRONMENTAL ASSESSMENT EXPERIENCE

The Department has completed a few SEAs, the most notable being the Canadian Environmental Reviews of the North American Free Trade Agreement (NAFTA) and Uruguay Round of Multilateral Trade Negotiations. However, except for these notable SEAs (and a few others) and some experience on project assessments related to properties abroad, the Department does not, at present, have a great deal of experience or expertise in conducting SEAs.

NUMBER OF POLICY AND PROGRAM OFFICERS

Given the broad mandate of the Department and the way the Department itself is structured, there are numerous sections responsible for preparing policies and drafting MC. In addition, the rotational nature of Departmental staff results in the frequent changing of staff. The result is that there is a large audience to "sensitize" and train.

CABINET CONFIDENTIALITY

The fact that most policies require Cabinet approval and are subject to Cabinet confidentiality, raises questions concerning the public consultation component. (This situation exists for all other federal Departments.)

TIMING

Often there is a requirement to respond to a policy issue within a limited time, which does not provide adequate time for an appropriate SEA.

WHAT ARE THE SOLUTIONS?

In order to meet these challenges, the SEA process had to complement the Department's existing MC approval process and convince employees of the benefits and requirements of conducting SEAs.

USE EXISTING SYSTEMS AND BUILD SLOWLY

The Department already had a process for reviewing and approving MCs. Senior management directed that this process be amended to include the requirements for SEA and to ensure that the Department's Environmental Services Division is contacted for each MC. As a result, the Environmental Services Division is better able to identify policies while they are still in the preliminary stages and can contact the relevant policy officers to discuss possible environmental requirements.

This first step went a long way in ensuring that all MCs originating from DFAIT, or coming to DFAIT for approval from other government departments, include a review of the environmental considerations when required.

However, to ensure that the environmental considerations were included in the development phase of the policies, it was recognized that a systematic and expanded process had to be developed. This process is based, to a certain extent, on the tools developed by the aforementioned Interdepartmental Working Group, but is designed to take into account the mandate and structure of the Department.

The Department has developed a two-phased approach. The first phase involves completing an *Environmental Implications Checklist* which indicates whether an assessment is required and if so assists in scoping out the requirements for the EA. The second phase consists of completing a more detailed EA if required.

Phase one: environmental implications checklist

The first section of the *Checklist* determines whether the policy requires an EA.

EAs are not required for the appointment of a senior official or the remuneration to an individual. There is also no requirement to do an EA for policies that respond to a clear and immediate emergency (e.g., disaster relief), national security, or cabinet process.

If an SEA is not required, the form is signed and filed for record purposes. Even though in many cases it is clear that no assessment is required, the form is signed and filed to reinforce to employees that the environment must always be considered. (It is anticipated that approximately 75% of all of DFAITs "official policies" fall into this category.)

If the policy is not excluded from the process, then the rest of the form must be completed. This section scopes out the EA by asking the following:

1) Is the policy or program to be considered by Cabinet or under the Minister's authorities?
2) Has an environmental review been completed for a similar proposal (if yes, provide reference)?
3) Have similar activities in the past resulted in environmental impacts?
4) Will someone else be considering environmental impacts as part of the program review?
5) Does the proposal directly involve or assist in the construction of infrastructure and thus triggers the Canadian Environmental Assessment Act?
6) Will there be a Regulatory Impact Analysis Statement prepared?
7) Will there be public consultation as part of the Policy and Program Analysis?
8) Do you think an environmental review is required?

If the policy does not require further assessment, the form is signed, filed for record purposes, and an appropriate paragraph is drafted to reflect these findings. At a minimum, MCs must contain a declaration that the policy or program under consideration has been assessed and that no environmental impacts are foreseen.

If further assessment is deemed appropriate, a detailed SEA must be completed.

Phase 2: detailed SEA

Although few in number, the Department does have responsibility in areas that could require a more detailed SEA (e.g., free trade agreements, multilateral agreements dealing with possible banned substances, certain funding agreements).

Policies requiring a thorough SEA will follow a recommended outline, which requests details on the proposal (including a description of the alternatives to the proposal and alternative methods of implementing the proposal), the methodology used, and an analysis on environmental considerations.

For each activity of the proposal, an analysis should be undertaken which covers:

- component of the proposal
- expected outcome
- possible interactions with the environment
- significance of the interaction and potential environmental impacts
- mitigation and monitoring to control or monitor potential negative environmental impacts

Additional information has been developed including Table 5.1, which shows some of the possible linkages between program changes, possible outcomes, interactions with the environment, and mitigation and monitoring. Also, information has been developed that provides advice specific to proposals that deal with trade, security, international assistance, sustainable development, and projecting Canadian culture and learning abroad.

Some of the global concerns that should be considered in the development of international policies or programs include:

- global warming from air emissions and loss of rain forest
- loss of biodiversity
- loss of habitant for wildlife
- soil erosion and formation of deserts from improper agricultural practices on marginal land
- degregation of water quality through the introduction of contaminants and untreated wastewater
- loss of wetlands and marine shoreline
- improper mining practices
- over-harvesting of renewable resources (i.e, beyond their ability to renew themselves)
- loss of current use of lands for traditional purposes

CONVINCE POLICY OFFICERS THAT STRATEGIC ENVIRONMENTAL ASSESSMENT IS IMPORTANT

To ensure that all policy officers accept the concept of SEA, it is important to point out to them that SEA is not considerably different from existing strategic planning methodologies, or existing planning and analysis processes. SEA is simply a struc-

TABLE 5.1

Program Leads to Changes In:	Possible Outcomes	Possible Interactions with the Environment	Strategic Mitigation for Monitoring
Economic Conditions	Changes in economic conditions, e.g., • competitiveness • available technologies • market conditions • trade relations/patterns • industrial structure (size, location) • resource use/ management practices • land use • transportation practices • regional development • urbanization • business practices	• loss/protection of wildlife, aquatic, marine habitat • loss/protection of soil productivity • degradation/ improvement of local air quality and water quality • atmospheric change through release of airborn contaminants, emissions	• regulatory processes and/or organizations in place • projected EA legislation or practice to be applied to physical processes • transfer of technology and environmental protection practices • educational/ training programs for the transfer of Canadian environmental protection practices • environmental status reporting and/or monitoring
Social Conditions	Changes in social conditions, e.g., • health conditions • population distribution • population demographics • work environment • recreation patterns • public, industry, or government consumption practices • individual awareness and behavior • concentration of changes in one group or population	• changes in land uses • changes in consumption of renewable resources (e.g., water, air, wildlife, fish, forestry) • changes in consumption of non-renewable resources (e.g., fossil fuels, mineral resources)	• environmental organizations such as commissions in place or to be formed • control of population migration • environmental management system and/or protection plans in place • historical record of environmental performance of organizations involved

TABLE 5.1 (CONTINUED)

Program Leads to Changes In:	Possible Outcomes	Possible Interactions with the Environment	Strategic Mitigation for Monitoring
Cultural Conditions	Changes in cultural conditions, e.g., • cultural/heritage resources • traditions or values • arts • religious and spiritual values and practices	• changes in land uses • changes in consumption of renewable resources • changes in consumption of non-renewable resources • loss/protection of cultural resources	• inclusion of environment clauses into the agreemen • application of Canadian environment of standards and technology to the proposal • incentive programs for improved environmental performance

tured process to consider and document the consideration of environmental and social factors early on in decision-making.

In many cases, policy officers have been examining the environmental aspects of policies for the last several years. Unfortunately these officers are either unaware that they are actually doing a SEA or, if they are aware, they have not documented it in a systematic format.

Policy officers need to understand that although the practice of SEA is still new and accountability mechanisms are still being developed, it is up to each employee to ensure that departmental policies and programs promote sustainable development. In doing so, they can be personally motivated because it:

- demonstrates an ability and commitment for sound policy analysis
- broadens ones professional development and experience
- improves the policy and avoids costly future remediation and restoration costs
- avoids delays with senior managements, who will demand environmental analysis prior to policy approval
- protects the environment by not damaging it through poor policy development
- ensures there are legal and government policy requirements for EA

Policy officers need to understand that the inclusion of environmental considerations in DFAIT's policy and program development achieves a number of Departmental objectives:

- ensures that sustainable development objectives are fulfilled
- demonstrates a commitment to the environment

- reduces environmental costs and improves health and social considerations
- improves efficiency and eliminates/lessens the need for remedial action
  at a project level
- expedites approval and project implementation
- avoids unnecessary delays for Cabinet approval
- protects the department from litigation under the CEAA

It is also crucial to remind Departmental staff that there are positive sustainable development consequences including:

- DFAIT's initiatives of foreign policy and international aid programs contributing to global sustainable practices
- contributions to global security with protection of the environment by
  avoiding damages associated with conflict
- foreign policies that prevent or mitigate armed conflict, which can reduce
  the destruction of natural resources
- trade rules and agreements that are a major determinant of how natural
  resources are used and/or exploited
- increase in environmental awareness in partner nations
- transfer of Canadian values
- transfer of Canadian technology, standards, and knowledge

CONCLUSIONS

The requirement for federal departments to conduct EAs of proposed policy and program initiatives is one component of a broad commitment by the government of Canada to reach its goal of achieving sustainable development. Other elements include legislation, regulations, and amendments to the Auditor General Act, and establishing in 1995 the Office of the Commissioner for the Environment and Sustainable Development. This Office will report annually to Parliament on the performance of departments in implementing their sustainable development strategies. The process will include goals, action plans, and the number and types of policy and program proposals that have been environmentally assessed.

The focus on EA represents a recognition that EA, which involves the identification and assessment of the environmental implications of proposals at all levels — policies, programs, plans, and projects, is a key tool in promoting sound decision-making in support of sustainable development. Implementing sustainable development requires decision-makers at all levels to have the capacity to make informed decisions. This means that they must be able to integrate economic, social, and environmental considerations into their decisions.

SEA provides a mechanism for ensuring that decision-makers account for the full range of impacts of their decisions. This involves giving equal consideration to economic, social, and environmental objectives; ensuring that trade-offs involving environmental impacts are explicit in promoting transparent decision-making; and taking reasonable measures to mitigate environmental impacts. SEA, properly applied and implemented, is not an "add-on" requirement, but an integral component of sound decision-making.

DFAIT is committed to meeting its sustainable development obligations by ensuring that it incorporates social and environmental considerations early enough to influence the design of policy. SEA is an important tool in reaching this goal. The Department will continue to refine its process and increase employee awareness. It is recognized that as this process is in its infancy, follow-up and monitoring combined with structured and one-on-one training is required.

The key lessons that emerge are: the value of simple, practical approaches; the need to begin the analysis as early as possible in the policy/program development process; the need to promote environmental benefits through SEA; and the need to inform decision-makers while there is still time to make a difference.

6 SEA in the Czech Republic

Milan Danny Machač, Vladimir Rimmel, and Lukáš Ženatý

CONTENTS

INTRODUCTION

The Czech Environmental Impact Assessment Act (No. 244/1992) took effect on July 1, 1992 and marked a step toward assessing proposals at the conceptual stage of planning. It is in stark contrast to the way EIA has been practiced over the last ten years.[1] While there are some shortcomings, it is hoped that the assessment at the strategic stage of program development will give decision-makers an opportunity to modify entire development programs.

The Czech EIA Act, § 14 Assessment of Concepts, provides formal framework for environmental assessment of policies, plans, and programms (ppp). Under this Act, a "concept" is a policy intention submitted and approved on the level of central state administration bodies in the field of energy, transport, agriculture, waste treatment, mining and processing of minerals, recreation, and tourism. Land-use planning documentation and the General Water Management Plan are also considered to be concepts.

Practical application of § 14 was limited before 1997 and experience with the implementation of Strategic Environmental Assessment (SEA) was heavily influenced by 40 years of communist rule, which produced a strong aversion of the Czech community to "plans," "planning," "strategies," and "concepts." Nearly no documents were produced or used of above titles and/or other titles. Sometimes the name of a document was changed in order to avoid assessment according to § 14.

Recently, the planners have begun to embrace § 14. During 1997, development of the state energy policy continued and works on transport and mineral natural sources policies commenced. All of these policies have been assessed according to § 14 (for details on SEA energy policy, see "What Works Well, and What Does Not?" on p. 85). Strategic assessment of three mentioned policies revealed short-comings of current legislation. Improving SEA process was one of the key compo-nents of the Czech EIA Act Amendment (a "draft").[2] The SEA part of the "draft" reflects most of the requirements of the "Commission Proposal for a Council Direc-tive on the assessment of the effects of certain plans and programs on the environ-ment," April, 1997.

The question of whether or not to include the land-use planning documentation into the SEA process has been discussed for several years. Recently, it was decided to assess only new plans and not half-done (unfinished) documentations.

There are two important reasons for the newly amended Czech EIA Act to be passed:

- to improve the current state of the EIA/SEA process
- to harmonize the Czech environmental legislation with the European legislation

STRATEGIC ENVIRONMENTAL ASSESSMENT IN THE CZECH REPUBLIC (WHAT ARE THE BIG ISSUES?)

Czech SEA experts and other SEA stakeholders expect that general key benefits of SEA implementation are:[3]

- incorporation of environmental considerations into all stages and certain sectors of policy-making
- promotion of integrated decision-making consistent with the approach outlined in Agenda 21
- clarification of environmental objectives, alternatives, and implications of development PPP
- identification of positive, environmentally friendly options and opportu-nities
- adoption of a range of impact mitigation measures
- modification of PPP proposals to take account of environmental consid-erations

Analyzing the current SEA situation and considering the three main categories (policy framework, institutional, procedural) of key SEA issues,[4] we can see that procedural issues seem to be most important in the Czech Republic. Appropriate policy and institutional framework are, of course, necessar,y too.

In the course of previous SEA development in the Czech Republic, policies, plans, and program, which might be subjects of SEA, have been identified. Principles and approaches to SEA process were specified in the draft of the Czech EIA Act Amendment and are very similar to those of the European Union (EU).

Czech developers have at each stage accepted the SEA requirements of certain PPP (specified in § 14 [2], [3], and [4]). For example, current strategic assessment of the state energy policy has brought about new options and higher quality of the policy.

Public involvement in the EIA/SEA process is still developing. The first experience with public participation at the SEA of energy policy is described in Chapter 4.

We are still at the very beginning regarding the linkage of policy formulation and implementation. We have been undertaking SEA process in regard to three important policies (energy, transport, and mineral and natural sources). In the case of energy policy, the process of policy formulation has been influenced already, and hopefully the others will be influenced as well.

During the second half of 1996, a research project, which contained SEA, was undertaken. Drafts of procedure and short-case study were elaborated. Using previous research and foreign experience, the Ministry of Environment is going to develop appropriate SEA methodology.

Three current SEA processes will be analyzed and evaluated by Czech and foreign experts, and consequently, appropriate methods and techniques for improvement and facilitation of the process will be developed.

Similar to the training of EIA procedure to all EIA stakeholders in the early 90's, SEA training to government officials, developers, policy-makers, and NGOs should be developed and delivered. We assume that several case studies and other methodological documents for SEA training will be developed using practical experience of three mentioned SEAs.

WHY DO WE PROMOTE STRATEGIC ENVIRONMENTAL ASSESSMENT?

Just like in other countries, the number of problems and obstructions regarding SEA process existed, so too did problems arise in the Czech Republic. There are many reasons. The past communist period has completely deformed the national economy. The new government (after 1989) had to change key principles and relationships in our society and economy immediately, and any discussions or assessments obstructed and delayed their intentions. Most of the concepts and long-term development plans from this period were, therefore, only political declarations. Each member of the state government (minister) enforced his or her idea of how to transform national economy. However, the level of communication and coordination inside and among the ministries was very poor. Consequently, the Czech government disputed each concept many times without any useful conclusions.

Hopefully this period of transition is over. Currently, an enormous effort is being made to complete the transition process and improve and accelerate the Czech economy growth. SEA could be one of the effective instruments for preventative environmental protection, and due to this promote and/or enforce the transition process. Last development shows that the competent ministries have understood (and/or have been forced to understand) that SEA application for assessment of key development policies and plans is an instrument which can prevent not only environmental damages, but bring positive economic and social effects.

Besides the necessity to overcome the above mentioned shortcomings, there are many other good reasons for promoting SEA implementation in the Czech Republic:

- The Czech Republic would like to be, as soon as possible, accepted as a new member state of the EU. A long-term process of approximation and harmonization of the Czech legislation with the EU legislation (including environmental) was started several years ago. The Czech EIA Act is nearly in line with the EU law. EIA legislation needs just some amendments to be in compliance with the new issues of Council Directive 97/11/EC and the proposal of SEA Directive (published April 25, 1997, *Official Journal of the EU* L 129). We believe the harmonization process has been, and will continue to be one of the key elements of SEA process promotion and development in our country. We expect promotion by the state and regional governments to continue. Such expectation is based on two main reasons. The first one is connected with obligatory assessment of certain PPP according to the amended EIA Act, and the second one is based on the belief that practical experience and benefits will convince developers as to the effectiveness of strategic assessment of their PPPs.
- Integrated Pollution Prevention and Control (IPPC) Directive is going to be implemented in the Czech Republic. The principles and requirements of IPPC should assist and promote SEA implementation and development. We suppose mutual influencing of SEA and IPPC processes. For example, if there are some development criteria defined within the industrial policy, individual developers will have easier roles for IPPC implementation. On the contrary, if there are no general criteria and principles, many more difficulties within the IPPC process can be expected.
- After six years (and about 1,000 EIA procedures) of practical experience with project EIA there is a lot of positive and negative experience, as well. In a number of cases, project EIA was a tool with certain limitations (see examples below). SEA seems to be a tool which can reduce or overcome these limitations.

EXAMPLES

The current case from the Ostrava region is an example of when EIA process could not cover and assess all circumstances and relations, however, SEA has not been used yet.

Industrial regions in the northeast of the Czech Republic need to modernize transport networks and connections with other parts of the republic and with neighbors Poland and Slovakia. Three major individual transport projects (Highway D47, highspeed railway, and the Oder, the Danube water channel) are under preparation. They are located in one corridor in a distance of only several kilometers from each other. EIA process of D47 took about two and a half years and the route of future highway was approved. However, the coordination and collaboration between the

Ministry of Environment (responsible for EIA) and the Ministry of Transport (responsible for transport policy) was poor. The alternative, to lead the highway in one corridor together with the railway and channel (according to the original EIS, it was the environmentally best alternative) was refused within the EIS review process. If the approved D47 route and the high-speed railway project are implemented, it is likely that approximately 6000 people will have to live on a new "artificial" island between the highway and railway. Plus, the approved route is about 3 km longer and more than 1 billion CZK (35 million U.S. dollars) more expensive.

If SEA process was implemented to assess possible environmental impacts (positive and negative) of the proposed transport network, basic principles, limitations, and requirements would be set down early enough. Following this, project EIAs should respect these principles. The result would be that the environmentally best highway route, coordinated with the proposed high-speed railway, would be approved. Significant economical savings would be reached as well. The EIA of D47 could also cover and respect other parts of the transport network.

Another example could be environmental assessment (EA) of a nuclear power station in Temelín. The assessment using the project EIA procedure would be inadequate and insufficient in relation to the current EIA Act and its time limitation. Sufficient answers to the necessity of this power station and the scope and significance of possible environmental impacts can be provided only with the strategic assessment of the energy policy.

WHAT WORKS WELL, AND WHAT DOES NOT?

The activities connected with the elaboration and assessment of the three policies (energy, transport, and mineral and natural resources) have brought about positive development in the field of SEA in the Czech Republic. It has been decided to assess environmental impacts according to § 14. Developers (in our case, the Ministry of Industry and Trade [MoIaT] and the Ministry of Transport) had to accept this requirement. Following is the characterization of key steps of strategic assessment of energy policy, which is most developed:

The MoIaT prepared and submitted in 1997 a proposal of the energy policy after a few years of preparatory works. The proposal involves a period following 10–15 years. We would like to stress an extraordinary importance on the energy policy in the postcommunist country.

Foundation SEVEn (Center for Effective Using of Energy) has been undertaking the SEA of the energy policy which was ordered by MoIaT. SEVEn has a lot of experience with environmental and energy projects. SEVEn is a non-governmental, non-profit consultancy founded in 1990. The goal of SEVEn is the protection of the environment and promotion of economical development using the more efficient energy utilization. For details, see www.svn.cz.

The Czech EIA Act has no obligatory procedure for strategic assessment. SEVEn had the difficult task of figuring out how to proceed, and they finally chose for a SEA statement elaboration, a procedure based on:

- systematic selection of basic impacts on the environment
- cooperation with the public
- use of the energy policy as a flexible material

The procedure which was set by SEVEn was consulted by foreign experts (Barry Sadler) and was discussed during the international conference EIA (Prague, 1998). All of the consultants confirmed that the procedure is appropriate and that its standard level is the same as in developed countries. They particularly emphasized significance of public participation.

The process of SEA was started in February 1998. The first step was the scoping process. The so-called Team A (advisory board of experts) was established and they prepared the first phase of the process, setting a scope of the assessment.

Six meetings of Team A were held, focusing on the creation of entering options even if it was not the original goal of Team A. It was a very important issue of the process because if only one option was developed and submitted, it might be a reason to stop and give the assessment process back to the Ministry of Environment (MoE) as a competent authority. The MoIaT compiled and submitted its draft of the energy policy with only one option. Team A developed the idea of the energy policy and specified three feasible options. Two of them resulted from the draft of the energy policy, and the third resulted in active public participation, particularly NGOs.

The options are as follows:

1. Development of an energy sector is based on domestic sources, such as black and brown coal. There are no limits on mining or export of fuel. It is hoped that electricity import will stop by the beginning of the year 2000. The construction of a nuclear power station in Temelín will be finished.
2. Development of an energy sector is based on domestic sources, as well. Mining of domestic sources is limited. The construction of a nuclear power station in Temelín will be finished. Import of black coal, including its products, will be limited, while other fuels will have no limits. There will be no limits for export, either.
3. Development of an energy sector is based on domestic sources, but there exists all ecological-town-planning limits for coal mining. The construction of a nuclear power station in Temelín will not be finished. There is an assumption for an improvement alternative and renewable sources of energy. There will be no limits for export of fuel.

Unlike a regular EIA process in the Czech Republic, the SEVEn has done its job in close cooperation with the public from the beginning. The number of NGOs remarks and recommendations was high and most of them brought helpful motions to the process.

The first public hearing was held in the first phase after a couple of meetings in the SEVEn. The following key issues were presented by Team A on this hearing:

- explanation of the term, "Energy Policy"
- discussion of options
- expected key environmental impacts
- setting of priorities of chosen basic indicators

The discussion was constructive and helpful for an executive organization (SEVEn). Most reproaches were focused on the general setup of the Energy Policy. The requirement of the third option was the most important issue which resulted from this public hearing.

The second public hearing was held in the Senate's Assembly Hall and was facilitated by Mr. Moldan, the Deputy Minister of Environment. The arising importance of the SEA process was confirmed by the participation of the representatives of Parliament Committees for Environment and by the Minister of the Environment. Fortunately, politicians didn't make use of the opportunity for political speeches, and instead discussed the perspectives of energy policy.

The internet is another way for the public to become involved and informed on the SEA process. All important information, partial results, dates of meetings, etc., were ventilated by the internet on the home page (www.svn.cz). It was a nice opportunity for the public to have discussions without time or distance restraints. The internet conference was mostly attended by regional non-government organizations located far from Prague.

Team B has been established at the end of the first phase. It has included experts who assessed chosen indicators in several variants. The base assessment includes simple assessment of indicators by each of the experts separately. The result includes two variants with/without economical indicators. Following this base assessment, SEVEn, using answers of several respondents, undertook another assessment. This assessment resulted in matching the weight with individual indicators.

Indicators were divided into four groups:

1. direct impact on environment (ecology)
2. impact on natural sources
3. social impact
4. economic impact

The importance of indicators were set according to informants' answers. Seven different informants have been selected according to their professional relations to environment or energy (ecological, sociological, NGOs, economic with relation to ecology, economic without relation to ecology, business with relation to ecology, and energy, including technicians with relations to ecology).

The last stage of public participation within the elaboration of the SEA statement of the Energy Policy were regional public hearings. Seven regional public hearings were held and facilitated by NGOs, which assisted the SEVEn, in this part of the process. It was the last chance for any comments. Public reaction varied significantly. Remarks included requirements on new alternatives (mostly a combination of three basic options), new indicators (or to leave out the older one), and questions regarding the choice of SEA methods. Of course, during public hearings, current government decisions regarding energy and environment were criticized. SEA statements have been finished and the next development might be:

- MoE elaborates an expert opinion, which will be submitted to the Government
- SEA statement will be opened to the public for 30 days at all district authorities
- remarks will be collected
- inter-ministerial commenting procedure will be done
- final negotiation will be made on the national government level

Besides worthwhile development of the above-mentioned SEA processes, other positive key aspects of SEA procedure in the Czech Republic can be summarized:

- satisfactory level of legislative compliance of the draft of Czech EIA Act Amendment with EU EIA legislation
- public interest on current SEAs is much more intensive now in comparison with previous five years
- nowadays, appropriate emphasis and interest of state authorities exists (see the public hearing of Energy Policy, which was held in the Senate)
- using internet for public involvement.

Negative key aspects of the EIA/SEA process are as follows:

- Appropriate SEA methodology is missing.
- The system of SEA training is missing.
- Enforcement of the SEA findings should be improved.
- Role of the public within SEA process should be formally (regulatory) determined.
- Political will is still unsatisfactory.
- Level of communication between ministries and inside the MoE particularly is weak.

WAYS FORWARD FOR STRATEGIC ENVIRONMENTAL ASSESSMENT IMPROVEMENT

The following are recommended steps and conditions for an effective SEA implementation in the Czech Republic:

- to complete and approve the EIA Act Amendment
- to develop and implement efficient enforcement tools
- to integrate SEA procedure with other related legal norms
- to implement IPPC Directive in the Czech Republic

- to charge the MoE (or other authority) with the responsibility for SEA and to create an effective system of communication within the competent authorities
- to determine appropriate roles for public participation
- to develop and deliver appropriate training for SEA stakeholders
- to determine relations between SEA and land-use planning

REFERENCES

1. Czech EIA Act No. 244/1992 Coll., 1992.
2. Draft of the Czech EIA Act Amendment, version October 1997.
3. Sadler, B., *International Study of the Effectiveness of Environmental Assessment*, Minister of Supply and Services Canada, 1996, 248.
4. Partidário, M.R., *Strategic Environmental Assessment-Highlighting Key Practical Issues Emerging from Recent Practice*, discussion paper presented in the International Association for Impact Assessment, Durban, South Africa, Annual Conference, 1995.

7 Evaluating Trade Agreements for Environmental Impacts: A Review and Analysis

William E. Schramm

CONTENTS

1-56670-360-3/00/$0.00+$.50
© 2000 by CRC Press LLC

INTRODUCTION: TRADE LIBERALIZATION AND THE ENVIRONMENT

International agreements deregulating trade have made the relationship between trade and the environment a matter of growing concern. Trade-environment interactions are complicated by the fact that they reflect not a single concern, but a diverse set of issues (e.g., social impacts, pollution spillovers, downward pressure on environmental standards, economic competitiveness, and loss of sovereignty) grouped under one heading.[1] This chapter examines trade-environment interactions in the context of trade liberalization and principles of environmental impact assessment (EIA).

There is strong evidence that liberalized trade supports expanded economic activity. The question of whether trade liberalization also supports environmentally sound and sustainable development is unresolved.[2] Trade liberalization affects nearly every aspect of society and the economy including: (a) the structure of economies (influencing what is produced, who will produce it, and where and how it is produced); (b) employment, national income, and the distribution of income within and among countries; (c) the rate at which, and the efficiency with which, renewable and non-renewable resources are exploited; (d) the rate of innovation and diffusion of new technologies; (e) the ability of nations to make investments in social and regional development; (f) the manner in which pollution standards are set; and (g) the mechanisms used to protect the global commons.[3] With numerous changes occurring simultaneously, the potential for synergistic effects (positive and negative) on social groups, the economy, and ecological resources is great.

Potential outcomes from the interaction of trade and the environment vary. Trade may permit some regions to benefit from development while "evading" environmental constraints (effectively importing environmental resources such as waste absorption from elsewhere), leaving other regions to bear the environmental and social costs. Alternatively, trade may, in theory, raise the level of sustainable development worldwide by compensating for limits to development, eliminating local imbalances, and contributing to the efficient use of resources.[4] Efforts to formalize protection of the environment in the context of global trade are not well advanced. Industrialized countries possess the institutional infrastructure to develop a consensus on environmental policies, allocate costs, and implement and enforce chosen policies. This type of infrastructure, however, is immature in less developed countries (LDCs) and is virtually absent at the international level.[5]

TRADE LIBERALIZATION

The Organization of Economic Co-Operation and Development (OECD)[6] lists four main types of trade agreements. *Sectoral agreements* control trade in a specific economic sector (e.g., the Multi-Fibre Agreement for textiles). *Commodity agreements* address trade in a specific commodity (e.g., the International Tropical Timber Agreement). Under *preferential trade agreements,* a country or countries (typically industrialized countries) offer preferred trade treatment on a non-reciprocal basis for imports from one or more countries (typically LDCs). *Trade liberalization agreements* involve changes in existing trade policies on a reciprocal basis within two or more countries with the intent of promoting unrestricted trade. Examples of trade liberalization agreements include the General Agreement on Tariffs and Trade (GATT) and the North American Free Trade Agreement (NAFTA). Trade liberalization agreements affect a larger set of economic sectors and commodities and, therefore, may have a wider range of environmental effects than other agreements.

During the last decade, the most significant market liberalization occurred in LDCs both voluntarily and under pressure from multilateral development banks or the International Monetary Fund. LDCs pushed industrialized countries to complete the Uruguay Round of GATT negotiations.[7] Two factors underlie LDC interest in less regulated trade. First, it is hoped that improved access to international markets will prove a more reliable source of income than foreign aid. Second, LDCs hope that trade liberalization through a multilateral entity such as the World Trade Organization (WTO) will produce a trading system based more upon rules rather than market power. Currently, countries or trading blocks with large, wealthy markets (e.g., the U.S. and E.C.) wield considerable coercive power over other nations because of the ability to restrict access to their markets.

Sources of Trade's Environmental Effects

The volume of world trade in 1990 was approximately $3.5 trillion (U.S.), about four times larger than in 1948 when GATT began. About 40% of trade occurs within multinational corporations.[8] Between 1948 and 1990, world energy use and gross domestic product (GDP) grew four-fold and world population doubled. With trade increasing, it is clear that a balance between trade-driven economic expansion and the environment cannot be sustained without substantial reductions in the environmental impact of economic activity. Such reductions, however, have yet to be observed.

Trade liberalization may have both positive and negative effects on the environment.[5,9] Trade affects ecosystems and human health through various mechanisms including: *scale* effects (i.e., more output means increased pollution and resource use), *technical* effects (i.e., the development and adoption of cleaner technologies), and *composition* effects. Composition reflects international specialization in cleaner or dirtier activities. With increasing development, the proportion of dirty production within an economy tends to decline relative to total production.[10]

Scale effects tend to increase pollution or resource use, while technical and composition effects typically work to reduce the pollution impact of production. Economic activity also impacts the environment through *landscape* effects: the specialization and homogenization of ecosystems[11] and change in the spatial arrangement of natural systems.[12] Trade liberalization also has a *regulatory* effect[9] (i.e., the impact on existing environmental policy and standards).

Environmental policies influence the nature of trade's environmental impacts. Appropriate policies support the realization of positive contributions from trade and mitigate negative impacts. Unfortunately, the general consensus is that, compared to the most developed countries, environmental policy in LDCs is less likely to be in place, appropriately designed, or enforced.[10]

EFFECTS AND CAUSAL LINKS

The environmental effects of trade are both direct and indirect. An illustration of a direct effect is transport. Increased trade brings greater activity in transportation and related sectors, including construction of transportation infrastructure, increased fuel use, and an increased frequency of transport-related accidents. Transportation support activities, such as the development and transport of fuel supplies, are also directly linked to increased transport demands.

The most significant environmental effects of trade, however, come from the indirect linkage of trade to economic growth, and economic growth to environmental quality.[12] Structural differences among economies mean that indirect environmental effects vary among nations. One important structural difference is the existence of dual-economies. The modern or industrial sector of the economy in LDCs exists side-by-side with a larger subsistence or informal sector (the susbsistence sector is very small or non-existent in developed countries). The size of the industrial sector varies across LDCs and may include manufacturing, mining, agricultural, forestry, or fishing. Operations in the modern sector are characterized by their relatively large size, high productivity, and relatively high wage scale. The subsistence sector may be rural or urban and is characterized by very small scale activity, low productivity, very low wages, and a lack of taxation and regulation.

Two levels of indirect environmental effects can be distinguished. The first is related to production activity in the industrial sector. The second is related to significant social change and has the most pronounced environmental impacts in the subsistence sector of LDC economies.

Environmental reviews of trade pacts have focused on the effects of production activity. When liberalized trade increases the level of economic activity, while products and production techniques are unchanged, the result is increased natural resource use, land use, and industrial pollution.

Other outcomes are possible, however. Greater efficiency and improved production and environmental technologies may be encouraged by liberalized trade and may partially or completely offset stresses generated by increased activity. Further, compared to a status quo of unilateral trade liberalization, trade pacts that result in environmental commitments and international cooperation that would not otherwise occur may reduce environmental stress in the medium to long-term. Where increased industrial sector activity does increase environmental stress, it does not automatically translate into damage to human and environmental health. Both human populations and ecosystems exhibit some tolerance of environmental stresses. This tolerance varies with the specific stressor and population subgroup or ecosystem type. Damaging levels of environmental stress tend to occur first where high-impact economic activities are located near sensitive populations and ecosystems.

While expanded production sector activity may increase employment opportunities and incomes, the short-term capacity of expansion to reduce LDC poverty and poverty-related environmental degradation is limited. Examinations of the effect of trade policy reform on welfare in small, open, distorted economies with environmental externalities (i.e., pollution)[13,14] indicate that the welfare effect of trade reform alone is ambiguous and varies with the pollution intensity of the sectors most affected. Given the lack of robust welfare gains from trade reform alone, the need to coordinate trade reform with environmental policy is emphasized.[13]

Environmental reviews of trade agreements have given little attention to indirect environmental effects stemming from social change, although they may be quite significant. The most dramatic examples of social effects appear in LDCs. The causal chain that relates social change to environmental effects tends to be more complex than that linking production activity to the environment. Further, the link between social change and the environment often involves poverty.

Trade liberalization advocates argue that freer trade improves the environment because trade leads to greater economic activity, greater economic activity increases wealth, and greater wealth leads to increased demand for a clean environment. This attitudinal (i.e., social) change, combined with a greater ability to pay for environmental protection, should lead to a cleaner environment.[15]

This scenario may hold for the industrialized sector, but it is not clear that it applies to the subsistence sector in LDCs. When previously protected markets are opened, there are both new opportunities and downward pressures on earnings and wages. Net income is determined by the competitiveness of each sector and economic gains are not guaranteed. Where trade does increase wealth, two caveats are necessary in assessing environmental impact. First, greater wealth does not guarantee that funds will be applied to environmental protection. Second, neither the benefits nor the costs of liberalization are evenly distributed within societies. Where LDCs have implemented economic and trade reforms as part of structural adjustment, economic benefits have tended to accrue to those in higher income groups (i.e., the modern sector) since they are in a position to capitalize upon new opportunities, while most costs have been shouldered by tenant and subsistence farmers, informal sector workers, and poor women[16] (i.e., the subsistence sector). For example, if trade liberalization leads large landholders in LDCs to switch to mechanized production, poor tenant farmers may be forced off the land, precipitating their migration to cities or settlement on undeveloped lands. Undeveloped land is typically available because it is environmentally fragile (e.g., hill slopes, floodplains, rainforest, dry lands) and unsuited for intensive development. The relationship between poverty and the degradation of fragile lands has been widely observed.[17-20] Since much of the environmental degradation in LDCs is poverty-linked, the distribution of winners and losers from trade liberalization is an important environmental concern.

Since the objective of trade reform is the maximization of overall welfare (rather than welfare from trade), it is appropriate and necessary to fully assess potential environmental impacts from trade pacts. The OECD[21] underscored the distinction between overall welfare and welfare from trade when it observed that "unlike sustainable development, free trade is not an end in itself."

MAKING USE OF EXPERIENCE

Many LDCs have undergone trade liberalization as a core feature of structural adjustment and stabilization programs. Their experience may, therefore, be of use in predicting environmental effects of trade liberalization. Conversely, approaches developed to analyze the environmental effects of trade liberalization may be of use when considering the effects of structural adjustment.

Structural adjustment programs promote export-oriented growth based upon a reduced economic role for the state, removal of market-distorting policies and structures, an opening to trade, and an increased reliance on competitive market discipline. These policies affect the mix of inputs to production, shift resources among industrial sectors and regions, and influence income distribution.

The experience of LDCs with structural adjustment indicates that in opening up to trade the extraction of natural resources for export tends to intensify.[16] Environmental damage has resulted, in part, because agencies that oversee natural resource development have been cut back at the same time that incentives for investment in resource extraction are increased. While reductions in the size of government are not specifically part of trade liberalization, they are a common component of economic liberalization measures prerequisite for participation in trade agreements.

The environmental implications of macroeconomic policy change have received little attention.[19,20,22] Evidence available, however, suggests that structural adjustment has produced mixed environmental impacts. The price corrections associated with structural adjustment programs (e.g., reductions in energy or water subsidies) have the potential to produce both environmental and economic benefits. These potential benefits, however, often go unrealized because price corrections are not accompanied by complementary policy and institutional reforms.[16]

ENVIRONMENTAL ASSESSMENT PRINCIPLES APPLIED TO TRADE AGREEMENTS

While environmental reviews of trade policy necessarily diverge from project level analyses in some respects, the basic assessment framework and principles are unchanged. The following sections consider elements of environmental review in the context of trade policy.

INITIATION

Initiation refers to the process of determining whether an environmental assessment is needed, the extent of analysis required, and how the resources and time available for the task are best applied. Since proposed trade agreements, environmental issues, and in-country institutional capabilities vary, no single assessment method can be specified. Mechanisms for determining the need for an assessment and its form include the review of lists of potential effects and the application of screening procedures. Combinations of these approaches are frequently utilized.

The main predictors of future environmental stress from trade liberalization are the current composition of economic activity, anticipated changes in economic activity (i.e., the composition of growth), current environmental policies and enforcement, and changes to environmental policy and enforcement as a result of a trade agreement. Within LDCs, the impact of trade pacts upon income distribution, the size of the subsistence sector, and social policies are also important predictors. If environmental problems exist under current circumstances, increased economic activity provides reason for concern. Environmental effects from existing economic activities are, in most cases, well recognized, although detailed characterizations may be lacking. With support from multi- and bi-lateral development agencies, country environmental reports have been developed for many nations. These reports may provide baseline information against which trade-related change can be measured. The nature of specific environmental impacts depends upon the type of activities within the economy undergoing reform (i.e., information/service, manufacturing, agriculture, or resource extraction). For example, in the manufacturing sector, consumption of a limited number of intermediate commodities explains more than 90% of variation in toxic industrial pollution in the U.S.[23] The use of these intermediates (Table 7.1) may indicate economic sectors most deserving of attention in assessments.

TABLE 7.1
Sectors Making Significant Contributions to Industrial Pollution

Sector	International Standard Industrial Classification	Products
Coal	(ISIC: 2100)	Hard coal, lignite
Oil and gas	(ISIC: 2200)	Crude petroleum, natural gas, gasified coal
Mining: ferrous	(ISIC: 2301)	Iron ores
Mining: non-ferrous	(ISIC: 2302)	Non-ferrous metal ores, uranium, and thorium ores
Mining: chemical	(ISIC: 2902)	Chemical and fertilizer minerals
Mining: other	(ISIC: 2909)	Peat, gypsum, andydrite, asbestos, mica, quartz, gem stones, abrasives, asphalt, other non-metallic minerals
Container paper	(ISIC: 3412)	Paper containers and boxes
Industrial chemicals	(ISIC: 3511)	Basic chemicals, nitric acid, ammonia, nitrate of potassium, urea, activated carbon, anti-freeze preparations, chemical products for industrial and laboratory use, nuclear fuels
Fertilizers	(ISIC: 3512)	Nitrogenous, phosphatic, and potassic fertilizers, pesticides
Paint	(ISIC: 3521)	Paints, varnishes, lacquers
Refining	(ISIC: 3530)	Refined petroleum
Pet-coal	(ISIC: 3540)	Miscellaneous products from petroleum and coal
Plastics	(ISIC: 3560)	Plastic products
Non-ferrous	(ISIC: 3720)	Non-ferrous metals
Construction	(ISIC: 5000)	Construction services, oil and gas extraction

Source: From "*Input-Based Pollution Estimates for Environmental Assessment in Developing Countries*," Dessus, S., Roland-Holst, D., and van der Mensbrugghe, D. OECD, Paris, 1994.

Given unchanged products and production techniques, increased production will increase environmental demands and the stress on environmental resources. Since trade-induced economic expansion or contraction will vary by sector, the current competitiveness of sectors is an indicator of future economic impacts. Countries entering trade liberalization agreements are likely to have undergone considerable liberalization already. Sectors that are doing well or poorly under current liberalization are likely to exhibit the greatest change in activity under intensified liberalization.

The degree to which trade and environmental issues are linked during trade negotiations varies. Environmental concerns were not included in the Uruguay Round of GATT. During NAFTA negotiations, the environmental effects of trade were addressed, but were segregated from trade talks. Currently, no country has a formal environmental screening mechanism for trade agreements. Canada probably comes closest; requiring an environmental review of all policy initiatives presented to the national cabinet. At a June 1993 Ministerial meeting, the OECD Council recommended that member states "examine or review trade and environmental policies and agreements with potentially significant effects on the other policy area early in their development." The OECD[6] developed examples of questions to use in screening trade agreements for environmental effects.

THE SCOPING PROCESS

Anticipated effects are examined during scoping to determine their importance. By bounding analyses, scoping makes the best use of human and financial resources and limits assessment costs. Scoping also identifies issues and data requirements that need specific and early attention. While scoping is an ongoing iterative process, it is most effective when begun early in assessment planning.

Scoping may be internal (i.e., government officials), or open to the public or select public subgroups (e.g., non-governmental organizations (NGOs)). Given time constraints of negotiations and the possibility of ongoing change to policies under review, internal scoping has real appeal for reviews of trade agreements. Experience with NAFTA, however, indicates that NGO involvement may raise issues, identify mitigation measures, and minimize public conflict.

The NAFTA experience also showed the value of integrating environmental discussions into the negotiation process.[24] The segregation of environmental issues into parallel discussion distinct from the main trade negotiations led groups in all three countries to appeal to national legislatures. Legislative pressure ultimately forced some environmental issues onto the main trade talk agenda, but the late introduction of these issues endangered and delayed the successful completion of the talks.[24] In the end, negotiations were slowed rather than accelerated by the parallel process.

Issues considered in scoping may be domestic, transboundary, or global. Domestic issues have received the greatest emphasis in environmental reviews of trade policy to date. These include the effect of agreements on domestic environmental law, air and water quality, renewable and non-renewable resource use, waste and toxic substance management, and wildlife management. When regional impacts were expected to be particularly significant, environmental effects on specific regions have received special attention (e.g., U.S.–Mexico border region during NAFTA review).

TREATMENT OF ALTERNATIVES

Alternatives considered during environmental review of a trade pact will vary with trade, economic, and environmental stress scenarios. Viewing change in a linear fashion, trade scenarios influence economic scenarios, which, in turn, affect environmental and social stress scenarios. These stresses combine with the distribution and susceptibility of human populations and ecosystems to produce environmental impacts. A cycle complete with multiple feedback loops, however, is a better representation of the real world (e.g., environmental effects from trade policy may feedback, through public pressure, to influence trade policy). Reviews also need to consider current environmental infrastructure (physical and institutional) and the treatment of environmental concerns in negotiations. With multiple scenarios for each variable, the number of hypothetical situations expands rapidly. Typically, a limited number of scenarios are used to limit the complexity of the situation matrix.

Trade scenarios include the sectoral, commodity, preferential, and liberalization agreements described earlier. Economic scenarios include the status quo (i.e., no trade agreement) as well as various growth scenarios. The status quo is typically an ongoing process of unilateral trade liberalization rather than no change. For the border region, the U.S. Trade Representative (USTR),[25] for example, postulated a growth scenario without NAFTA and two scenarios for NAFTA-induced economic growth.

ANALYSIS OF POTENTIAL IMPACTS

Environmental assessments (EA) may be organized by resource area (e.g., ecology, air, water, land, socioeconomic, cultural) or by sectoral divisions (e.g., transportation, commerce, natural resources). The latter may be particularly suited to analyses of trade agreements, given the role that economic data may play in reviews. Analyses of economic scenarios could use econometric models to a greater extent than is typical of project assessments. Similarly, environmental reviews of trade agreements will require more analysis of environmental regulations and institutional factors than is seen at the project level. The most striking variation from project analysis, however, may be the amount of scenario analysis involved. In the demonstration of a qualitative approach to assessing environmental effects of trade agreements on LDCs, Harwell et al.[26] considered 3 trade scenarios, 20 economic sectors, 2 stress characteristics for each of 40 environmental stresses, 23 ecosystem types, and 2 levels of environmental regulation. This resulted in more than 200,000 scenario combinations.

Trade policy change produces socio-economic change in a setting of ongoing socio-economic change. Thus, synergistic effects may be important. For example, given a situation in which trade liberalization has little effect on the forestry sector, but a large effect on subsistence agriculture, logging may still be a critical factor in facilitating land degradation by enabling poverty-driven migration to, and settlement within, previously inaccessible areas. Consequently, an emphasis on indirect and cumulative effects analysis should distinguish policy-level review from project analysis.

Whether analyses of potential environmental impacts are quantitative or qualitative, environmental reviews will share certain basic characteristics. These include:

identifying possible environmental effects, defining a baseline set of environmental conditions, determining the availability of data and appropriate means of analyzing the data, and estimating impacts and related uncertainties.

Use of Quantitative Data and Models

Models to provide policy relevant information on trade-environment interactions are under development. The Commission for Environmental Cooperation (CEC)[27] reports on macroeconomic, environmental assessment, and combination models. Macroeconomic models attempt to capture interactions that result from economy-wide policy change. These interactions include indirect economic effects resulting from changes in income, savings, and investment; resource competition among sectors; and producer-consumer interactions. Criticisms of these models relate to the assumptions involved, the use of broad sectoral categories, limited geographical coverage, and the absence of environmental variables.

EA models identify relationships between economic change and environmental impacts, although they do not typically address trade liberalization directly. While these models incorporate a much broader set of environmental information than do macroeconomic models, gaps in environmental information remain a major obstacle. The models are typically developed for specific environmental concerns such as global warming or acid rain.

Combined models link portions of macroeconomic models (e.g., trade analysis components) and models addressing environmental quality. Currently, these efforts focus on production effects (e.g., industrial pollution) and do not capture poverty-related environmental degradation.

Obstacles for all models are the availability of appropriate environmental data, and uncertainty about links between trade liberalization and the environment. Current shortcomings limit modeling's suitability as an aid to policy formulation.[27] Further, quantitative assessments involving significant data collection or modeling efforts may place unmanageable demands on the resources of some countries. To date, assessments of trade agreements have been qualitative rather than quantitative.

Qualitative Analyses

Assessments using expert opinion can provide policy-makers with information on the areas and groups likely subject to environmental impacts and can suggest mitigation measures. Harwell et al.[26] describe a method that uses expert opinion to analyze environmental effects of trade agreements in LDCs. Its focus is on effects from production activities; poverty-related effects were not considered. The method identifies vulnerabilities (i.e., portions of ecosystems exposed to risk from trade-induced growth) and evaluates the risk-reduction achieved by implementing additional environmental controls. For each trade/economic growth/environmental stress scenario, two outcome levels are sought. The first is the environmental effect based upon current environmental policy and enforcement, and the second is the environmental effect under more stringent environmental controls.

Harwell et al. considered effects in Venezuela from a hypothetical trade agreement with the U.S. The geographic distribution of current economic activity and potential economic changes from a trade pact were used to produce a spatial distribution of environmental stress. Stress levels and the distribution of ecosystems were used to develop estimates of potential environmental effects. The product was an ordinal ranking of ecosystems at risk and the industries most responsible.

Analysis of Global Concerns

In EAs of trade agreements, transboundary pollution concerns have been the primary international focus. The evaluation of global impacts from proposed trade agreements pose particular problems. Since no single country or organization has responsibility for global impacts, there is a considerable risk that these issues will receive inadequate attention. Potential global impacts can be divided into two categories: truly global impacts and impacts on the global commons. Global impacts include global warming and ozone shield degradation. Here trade liberalization's effect may be both positive and negative. For example, increasing industrial and transportation activities would increase greenhouse gas emissions. Conversely, trade agreements could accelerate conversion to cleaner energy technologies, reducing emissions per unit of production.

A major global commons concern is ocean resources. The U.S.–Mexico tuna controversy is an example. The U.S. unilaterally imposed restrictions on Mexican tuna imports because safeguards to prevent the capture of dolphins by the Mexican tuna fleet were deemed inadequate.[28] Canada[29] summarizes the view of many nations that disagreed with the U.S. action. With the exception of measures based upon broad international consensus, Canada rejects extraterritoriality because it undermines the sovereign right of nations to manage their internal affairs. Large, less trade-dependent nations may use trade restrictions to gain undue influence over nations more dependent upon trade (e.g., exports represent 7% of the U.S. economy and 24% of the Canadian economy). Canada also fears that protectionist lobbies in other countries would use environmental concerns in order to advance their own commercial interests. Nations supporting Mexico in the dispute felt the U.S. action had more to do with the protection of U.S. fishermen than dolphins. A GATT trade panel ruled for Mexico, agreeing that absent a multilateral environmental agreement (MEA), a country may not use trade restrictions as a means of protecting environmental resources beyond its borders.

Controversy extends to the definition of a global issue. Some international concerns (e.g., biodiversity loss, land degradation) involve resources that are different in a political sense from "unowned" atmospheric or oceanic resources. Rainforest, for example, is owned by sovereign states that often object to the "global resource" label and its implication that other nations have a "right" to involve themselves in what are seen as domestic decisions concerning natural resource use.

Environmental effects in one country are a legitimate subject for environmental analysis by a second country when actions of the second country (e.g., entering a trade agreement) contribute to impacts in the first. As a practical matter, however, it is difficult for one country to analyze activities occurring in another. Questions of national sovereignty and access to data limit both the practicality and desirability of such analyses. Consequently, there are advantages to each party to a trade agreement undertaking an independent environmental review. This approach was used with NAFTA. No organization currently has authority to organize the preparation of international environmental reports on agreements such as GATT. On March 15, 1999, the Director General of the WTO called for the creation of a World Environmental Organization parallel to the WTO in order to establish a multilateral rules-based system for the environment. Presumably such an organization would have authority to organize environmental reviews on international agreements.

OUTSIDE REVIEW AND PUBLIC PARTICIPATION

The objective of outside review is to ensure that assessments are complete and conclusions are well-reasoned and defensible. This is achieved through examinations of assessments by outside experts and organizations, and requiring full consideration of review comments. Preparers must defend analyses in light of comments or change both the analyses and conclusions based upon them.

Public participation refers to the involvement of the general public (individually or as members of organizations) in the environmental review process. Public participation is one of the more controversial aspects of environmental review. In many societies, no tradition of providing opportunities for public input to government decision-making exists. In these instances, such public participation may serve as a tool in democratization efforts. Even in societies where public participation is supported by tradition and law, there may be resistance from government officials. This recalcitrance is due to the perception that public participation is time-consuming and a source of problems. Public involvement, however, surfaces existing concerns rather than creating them. By uncovering issues early, it provides time to address issues or, at a minimum, allow full public discussion. Frequently, an open public discussion increases support for an agreement whether or not a particular issue is satisfactorily resolved. The experience of NAFTA illustrates the importance of public involvement, at least by NGOs. NGOs were given the opportunity to review the draft U.S. NAFTA environmental report and did so extensively. These comments played an important role in shaping NAFTA's environmental side agreement and enhanced acceptance of the overall NAFTA agreement. This was achieved without compromising the ability of the parties to negotiate.[24]

With ongoing trade negotiations, time issues regarding public input are real. Official comment periods do not fit well with efforts to respond to changing policy options during talks. If environmental assessments are initiated only after negotiations have produced a final policy proposal, however, the opportunity to influence negotiations has been lost. Canada addressed this problem during NAFTA negotiations by giving one set of reviewers responsibility for EA and for ensuring that, throughout negotiations, negotiators were aware of environmental consequences of policy options.

MONITORING AND FOLLOWUP

In project-specific EA, monitoring refers to measures undertaken to ensure that impacts do not exceed predictions and environmental commitments made as part of the assessment process are fulfilled. For trade pacts, the concept is extended to broader mechanisms, such as boards created to respond to environmental problems and provide for continuing public input. Trade agreements could provide for evaluations of environmental, economic, and social factors following implementation; including the production of periodic, publicly available reports. The CEC, which grew out of the NAFTA environmental side agreement, is currently engaged in a multi-phased effort to assess post-NAFTA environmental effects. CEC activities represent an opportunity to advance state-of-the-art EAs for trade agreements. These efforts could: enhance assessment methods; refine data collection methods and improve data availability; strengthen the analytical basis of modeling assumptions and the models themselves; and test the validity of specific trade-environment linkages and the effectiveness of various mitigations. NAFTA's adoption of follow-up studies supports the inclusion of similar follow-up provisions in future trade agreements.

ENVIRONMENTAL REVIEWS OF TRADE AGREEMENTS

For the GATT Uruguay Round agreement, the U.S. appears to have been the only nation to have performed a specific review. Neither the European Community, individual European nations, or Japan prepared environmental reviews of GATT. For NAFTA, each signatory prepared an environmental review (the U.S. prepared two). The two U.S. reviews and the Canadian review are addressed below. The Mexican environmental review of NAFTA was never publicly released.[24]

THE GENERAL AGREEMENT ON TARIFFS AND TRADE

The GATT grew out of a post-war effort to prevent a recurrence of trade conditions that contributed to economic and political instability in the 1930s and the onset of World War II. The architects of the post-war economic system recognized that trade liberalization was continually at risk from political failure.[1] They viewed the downward economic spiral leading to World War II as the product of domestic politics that pushed national policy toward non-cooperation on trade.

Those designing post-war policies concluded that the solution was an international regime to reinforce political structures in the face of domestic pressures. GATT embodies this solution. It supports collective action intended to benefit all by helping each nation resist policies based upon narrow self-interest. In this, reciprocity is central. By accepting modest restraints on sovereign power, each nation receives guarantees of a significant measure of respect for its interests in foreign markets.

Difficulty getting environmental issues on the GATT agenda results from GATT's narrow focus and insulation from political pressure. These characteristics were intentionally designed into the organization to help GATT withstand protectionist demands and challenges from special interests.[1] At the time that GATT was established, environmental issues were simply not on the agenda.

Environmental Elements of the GATT

The environment is not specifically mentioned in the GATT General Agreement text. While a number of GATT provisions may apply to the environment, GATT's interaction with the environment is a function of the manner in which these environmentally related provisions are interpreted.

Environmental concerns about the GATT can be reduced to criticisms involving two areas: its procedures and substantive rules.[1] The general criticism of GATT procedures is that they reflect a systematic bias toward trade concerns. GATT's substantive rules are criticized because they are narrowly focused on the commercial benefits of trade and do not reflect environmental considerations. GATT procedures do not require an analysis of possible environmental effects of trade liberalization. While good trade policy-making requires knowledge of related environmental issues, GATT trade negotiations are allowed to proceed while "in the dark" environmentally.[1] GATT's dispute resolution process receives more criticism than any other aspect of GATT operations with regard to the environment. The process is asymmetrical. GATT reviews cases where it is argued that environmental standards are too high and may rule that they represent unacceptable barriers to trade. No mechanism exists, however, to determine that environmental standards are too low and create unfair trade advantages based on cost externalization.[1] GATT puts the burden-of-proof on nations accused of having deviated from GATT rules rather than on nations claiming their trade position has been impaired.* The result is that under GATT, environmental rules are guilty until proven innocent.

GATT rules allow countries great flexibility in regulating products sold in domestic markets, but provide no mechanism for trade actions in response to cross boundary pollution or degradation of the global commons. Exceptions to GATT rules are allowed in accordance with Article XX. While it is possible to interpret Article XX quite broadly, the current interpretation based upon decisions of dispute resolution panels is quite narrow.[1] In evaluating the "necessary" test, dispute resolution panels have "read into" GATT a requirement that environmental regulation, in order to be GATT consistent, must be the "least GATT-inconsistent" policy tool available.[1,2] This interpretation poses a high hurdle for environmental regulation. GATT rules have been interpreted as allowing regulation of environmental variables related to products, but not production processes. The distinction between the environmental effects of products and production processes, however, is artificial, difficult to apply as industry moves toward life cycle assessment of products, and irrational in an ecologically interdependent world where pollution spillovers from production processes may have global consequences.[1,2] Further, GATT's product verses production

---

* Under NAFTA, the burden-of-proof lies with the country bringing a challenge.

process distinction has already been abandoned in some contexts. GATT regulates production methods in the code on subsidies and the Uruguay Round provision on intellectual property.[1]

Uruguay Round Changes

The environment was not on the agenda for the Uruguay Round of GATT talks. Thus, the Round did little to change provisions related to the environment. Unchanged provisions include: the text and interpretation of Article XX as it pertains to the environment, the relationship between GATT and MEAs, GATT rules regarding the use of trade measures for environmental protection purposes, and rules on discrimination among products based on production processes. Some changes that occurred during the Uruguay Round are supportive of environmental protection; others are expected to make environmental protection more difficult. The Uruguay Round established a new organization, the WTO. The preamble of the agreement forming the WTO explicitly commits the organization to the objective of sustainable development. This formal commitment to sustainability provides a basis for member nations to push for greater consideration of environmental concerns.

An agreement on sanitary and phytosanitary measures (S&P) refined the Article XX as a "necessary" test.[1] The language states that standards should be no more trade restrictive than "required" and that measures are GATT inconsistent only if "there is another measure reasonably available taking into account technical and economic feasibility, that achieves the appropriate level of protection and is significantly less restrictive to trade." The terms "reasonably available" and "significantly" have the effect of lowering the hurdle that environmental standards face in meeting the "necessary" test.

The Uruguay Round made it clear that nations have the right to choose levels of environmental risk they deem appropriate without fear of challenge under GATT. Other changes should increase the transparency of the dispute resolution process. For example, countries may now choose to make public their submissions to dispute resolution panels and, upon request, must release non-confidential summaries of panel submissions. Additionally, dispute panels will be permitted to call upon environmental experts when considering environmental issues. The parties to a dispute, however, do not have the right to demand that a panel seek assistance from experts. The Uruguay Round also specified a set of conditions under which environmental subsidies are permissible (i.e., not subject to challenge) and formed a WTO Committee on Trade and Environment.

On the other hand, the new Technical Barriers to Trade (TBT) discipline requires that environmental regulations not be more trade-restrictive than "necessary." Dispute resolution panels have in the past interpreted this language as meaning that environmental regulations must be the "least-trade restrictive" available in order to be consistent with GATT. If this traditional interpretation is applied, the TBT agreement would formalize in the GATT text the high hurdle that environmental regulations have previously been subjected to in dispute panel decisions. The TBT agreement also requires governments to revise national regulations if circumstances or environmental goals change and the changed circumstances or objectives can be addressed in a less trade-restrictive manner.

U.S. Review of the GATT

The Office of the USTR prepared a report on environmental issues related to the GATT Uruguay Round Agreements.[30] This review, however, was not intended to meet the standards of U.S. law. If the purpose of environmental review during trade negotiations is to highlight environment-trade issues and provide negotiators with a broader context for policy discussions, then the USTR Environmental Report was deficient in important areas. The report was not produced in time to be used by negotiators. On March 1, 1994, the USTR issued a notice requesting public input on the review's scope. The Uruguay Round Agreements, however, were concluded on December 15, 1993 and signed on March 15, 1994. Thus, the environmental review occurred after Uruguay Round Agreements were finalized and after changes could be made in the text.

The report did not systematically consider the potential environmental effects of alternative scenarios (e.g., various rates of U.S. economic growth, potential shifts in resource use among sectors, or the effect of not implementing the agreements). The focus was on economic and environmental effects within the U.S. and the effect of the agreements on U.S. policy, laws, and standards. In this context, the report provided background on the Uruguay Round proposals and discussion of potential impacts of public concern (e.g., effects on environmental standards and commercial interests).

Although the report noted effects on a sector by sector basis, it did not address all major sectors. In a number of cases, the economic effects on a sector received more attention than the potential environmental effects. Further, the consideration of environmental effects often consisted of a listing of potential impacts with no real effort to assess scale. An evaluation of environmental change was complicated by the fact that no baseline set of environmental conditions was defined.

With regard to macroeconomic effects and their environmental impacts, the report tended to focus on "brown" (i.e., industrial pollution) issues. "Green" issues (e.g., land and habitat degradation) received limited attention. The report also tended to presume that environmental regulations would prevent significant environmental impacts. EIA practitioners, however, have long been aware that regulatory protection of environmental resources (even when effective) does not eliminate the potential for environmental impacts. A discussion of the strengths and weaknesses of the U.S. regulatory framework in relation to anticipated environmental changes would have been useful.

A benefit of EA is that it provides an opportunity to devise approaches that mitigate environmental impacts. Although the report described various programs and policies that would counter potential negative environmental impacts related to liberalization under the Uruguay Round, it made no specific recommendations concerning mitigation. The USTR sought public input on the scope of the report before preparation, but not on the final report. Overall, the "tone" was one of "defending" or "selling" GATT, rather than systematically evaluating environmental issues and public concerns. This, and flaws in approach, undermined confidence that the analysis was unbiased.

THE NORTH AMERICAN FREE TRADE AGREEMENT

The U.S.–Canada Free Trade Agreement was implemented without environmental provisions or review. In 1990, however, when the U.S., Canada, and Mexico announced their intention to negotiate a trade agreement, environmental concerns became a major factor in negotiations and in the approval process. Negotiations were initiated in June 1991 and an agreement was signed in August 1992. In the fall of 1992, presidential candidate Bill Clinton committed to negotiations on an environmental side agreement, North American Agreement on Environmental Cooperation (NAAEC). Following Clinton's election as U.S. president, negotiations on NAAEC were initiated. The final text of NAAEC was released in September 1993. NAFTA became effective January 1, 1994.

Environmental Elements of NAFTA

NAFTA contains at least two provisions addressing environmental concerns.[31] The chapters on S&P measures and on standards recognize the right of signatories to set levels of protection they consider appropriate. While consistency in standards is urged, there is no "least restrictive to trade" stipulation as in GATT, and no more than some scientific basis for a standard is required (a lower standard than GATT). These provisions allow signatories to maintain existing standards although questions remain about higher standards imposed by non-participants (e.g., U.S. states) and the precedence of existing environmental treaties. Regarding the latter, NAFTA explicitly recognizes precedence of several multilateral agreements (e.g., the Convention on International Trade in Endangered Species, the Montreal Protocol, and the Basel Convention).

The parties also negotiated an environmental side agreement, NAAEC, which commits signatories to: a) enforce environmental laws; assess NAFTA impacts; report on the environment; develop emergency measures; and support education, research, and development; and b) establish a commission, the CEC, to coordinate implementation of NAAEC commitments and resolve disputes. The agreement also established a Border Environment Cooperation Commission and a funding mechanism (North American Development Bank) to address U.S.–Mexico border pollution concerns.

U.S. Review of NAFTA

The USTR prepared two reports, February 1992[25] and November 1993,[32] on environmental issues related to the NAFTA agreement. The 1992 report (issued as a draft in October 1991) was prepared concurrently with NAFTA trade talks, provided a number of recommendations to negotiators, and influenced U.S. positions in the NAFTA discussions and subsequent talks on the environmental side agreement. The focus was on issues related to U.S. standards and environmental concerns in the U.S.–Mexico border region (e.g., air and water quality, hazardous and municipal waste, chemical emergencies, transportation, demographic effects, wildlife and endangered species, and investment). The USTR sought and received public input on the 1992 report. This input influenced the final environmental review, and both NAFTA and NAAEC talks.

The 1992 Environmental Report discussed three growth scenarios for the border region: growth at the same accelerated rate as the rest of Mexico under NAFTA (high growth), a relative decline in border growth from a NAFTA-induced, southerly shift in the siting of new Mexican industrial facilities (slow growth), and growth at absent NAFTA (baseline growth). The range of growth estimates and the overlap among scenarios was great (base growth rate: 5–15%; high growth: 6–17%; low growth: 4–12%) indicating considerable uncertainty about variables influencing border growth.

The 1993 report was more comprehensive than the earlier report, addressing macroeconomic effects, environmental impacts on the entire border region, some effects beyond the border[24] and NAFTA and no-NAFTA scenarios. Product standards were the primary topic not specifically related to the border. Since it was produced following the conclusion of both NAFTA and NAAEC talks, however, it influenced neither agreement. Rather, the 1993 report evaluated developments occurring after the 1992 review including the adoption of specific environmental provisions within NAFTA and NAAEC, and it updated analyses performed earlier. Neither review meets the standards of U.S. law.

Both reports concluded that NAFTA's potential environmental benefits were significant while adverse effects were limited. The reports anticipated improvement in Mexico's environmental situation with NAFTA because: a) economic growth would increase the resources available for environmental protection and domestic demand for improved environmental quality, b) development pressures in the border region would ease with increased incentives for siting industrial facilities in the interior, and c) access to clean fuels and environmental technology would improve. Additionally, the reports concluded that Mexico would not serve as a "pollution haven" for industry.

Both reports focused on economic and environmental effects in the U.S. and the relationship between NAFTA and policy, laws, and standards. The reports provided background on NAFTA proposals, discussed potential negative impacts on U.S. environmental standards and commercial interests, and evaluated how NAFTA did, or did not, affect economic and environmental concerns.

The set of economic sectors considered by the reports was not comprehensive. Additionally, the reports were more effective in linking trade changes to economic effects than in linking economic effects to environmental impacts, and they focused on pollution issues to a greater extent than land and habitat degradation. Environmental effects were treated broadly, frequently listing potential impacts without qualifying scale or intensity. In most cases, environmental impacts were not linked to specific sectors and there was a presumption that existing regulation would prevent significant impacts.

With the exception of border population growth and a prediction of higher "average" incomes for all signatories, social impacts were not addressed. Mexican environmental effects without the potential for direct spillover into the U.S. received limited consideration (e.g., Mexican agriculture and forestry were discussed only briefly). The reports did not address NAFTA's impact on global issues.

The reports discussed the role of intergovernmental cooperation and U.S. regulations in responding to environmental and economic changes. They did not, however, evaluate potential weaknesses in U.S. environmental regulations or suggest changes to mitigate environmental effects.

Canadian Review of NAFTA

The Canadian review had two goals: ensuring that environmental objectives were considered throughout negotiations, and preparing an EA. The Canadian review process included regular meetings of the environmental review and negotiation teams, continuous review of the evolving agreement by the environmental team, and opportunities for public input. The Canadian review was structured as a programmatic EA and characterized environmental concerns generally. Environmental screening addressed global and atmospheric issues, air and water issues, sectoral resource issues (agriculture, fisheries, forests, wildlife, protected areas, and energy), and the management of toxic substances and waste. Other issues included environmental standards, investment, industry migration, extraterritoriality, and follow-up mechanisms.

The report focused on economic and environmental effects in Canada and evaluated NAFTA and no-NAFTA scenarios. The report concluded that NAFTA had little effect on the Canada–U.S. relationship and that a ten-fold increase in Canadian exports to Mexico would not have a significant effect on Canada's environment. Given this, the report did not identify socioeconomic impacts.

The report addressed effects on Canada's major economic sectors. It acknowledged difficulty developing quantitative impact estimates, in part, because the economic effects were expected to be small. The report implicitly defined its baseline as the time of report preparation (i.e., 1992) and provided reasoned arguments to support its conclusion that environmental effects would be nominal. The report balanced consideration of "brown" and "green" issues more effectively than did the U.S. reports, but like the U.S. reports, it did not discuss the strengths and weaknesses of Canada's regulatory framework in relation to anticipated changes, and it made no recommendations concerning mitigation.

CONCLUSIONS

Trade liberalization does not directly cause environmental impacts. It does, however, induce economic and social change that may create new environmental stresses and aggravate existing ones. To prevent environmental degradation and contribute to the development of complementary trade and environmental policy, environmental reviews of trade agreements must be useful (i.e., effects must be predicted with reasonable accuracy) and it must influence the formation of trade policy.

Prediction accuracy derives from an understanding of how trade liberalization-induced change contributes to environmental stress and how environmental impacts flow from environmental stresses and the vulnerability of specific populations and ecosystems. Spatial aspects of factors are critical.

Current economic and environmental conditions may indicate future vulnerabilities since increased trade has a limited effect on economic structure in the short to medium term. Growth in existing sectors is a more immediate response to trade growth than is the start of new activities. The status quo, absent a trade pact, is often a process of unilateral liberalization rather than no change.

Both positive and negative environmental effects are possible from economic shifts. Expansion of economic activity may result in more pollution and resource use. Conversely, trade liberalization may strengthen incentives for efficient resource use (e.g., removing subsidies that contribute to resource waste). Realizing environmental gains from price corrections, however, is not automatic. Potential benefits of macroeconomic policy reform often go unrealized because price corrections are not accompanied by complementary policy and institutional reforms.[16]

Where the economic impact of trade policy change is expected to be strong, that sector should receive special emphasis. In the U.S. economy, a small number of economic sectors make significant contributions to toxic releases (Table 7.1). Where these sectors represent a significant part of an economy now, the impact of trade liberalization upon their operations should receive close attention.

The intensity of environmental stresses on ecosystems is influenced by the effectiveness of mitigation, which, in turn, depends upon the strength of environmental legislation and enforcement activities. Where mitigation is currently inadequate, trade-induced economic expansion will exacerbate the situation absent specific preventative action. The capacity of more stringent "traditional" environmental regulation to reduce trade-related environmental impacts, however, may vary with the type of environmental stress involved. Current regulatory forms address pollution issues more effectively than issues of habitat and species protection. More stringent traditional environmental regulation may reduce the growth of industrial pollution from economic expansion without much effect on land and habitat degradation.[26] Innovative means of protecting environmental resources should be considered within environmental reviews of trade agreements.

Where poverty-related environmental impacts occur currently, trade-induced social change may exacerbate these impacts. Economic pressure upon the subsistence agriculture sector can lead to sudden and dramatic environmental degradation. For this reason, the distribution of economic "winners" and "losers" from trade policy change may serve as an indicator of environmental effects.

Environmental reviews of trade agreements to date have considered: a) the effect of trade pacts on current environmental regulation and future environmental policy-making, b) environmental impacts on border regions, and c) effects from macroeconomic change. This subject order reflects both increasing difficulty and decreasing effectiveness of review. Conventional pollution has been the primary environmental impact addressed. Land degradation from commercial activity has been a secondary consideration. Poverty-related environmental degradation has received little attention.

The likelihood that trade liberalization is accompanied by complementary environmental and institutional policies increases when environmental concerns are included within trade talks. For this reason and because trade pacts and associated environmental agreements may modify environmental protection regimes, reviews should consider how trade pacts address the environment.

Reviews of trade agreements indicate that the negotiation process is amenable to the incorporation of mitigations suggested by environmental review. Liberalization agreements themselves, or environmental agreements tied to them, may provide effective vehicles to address the environmental stresses created by expanded trade. To influence trade policy, however, environmental review must occur during the negotiation process, and this has not always been the case.

Finally, public interest and pressure are important in ensuring the quality of environmental review. Public interest was greater during the U.S. review of NAFTA than during the GATT review. The result was that the NAFTA reviews were more thorough and more effective in influencing policy.

REFERENCES

1. Esty, D., *Greening the GATT: Trade, Environment and the Future*, Institute for International Economics, Washington, D.C., 1994.
2. Repetto, R., *Trade and Sustainable Development*, United Nations Environmental Program (UNEP), Geneva, Switzerland, 1994.
3. Runnels, D. and Cosbey, A., *Trade and Sustainable Development — A Survey of the Issues and A New Research Agenda*, International Institute for Sustainable Development, Winnipeg, Canada, 1992.
4. Dahl, D., Global sustainability and its implications for trade, in *GATT Symposium on Trade, Environment and Sustainable Development, July 28, 1994 (TE009)*, General Agreement on Tariffs and Trade (GATT), Geneva, Switzerland, 1994.
5. Anderson, K. and Blackhurst, R., *The Greening of World Trade Issues*, Harvester Wheatsheaf, New York, 1992.
6. OECD, *Methodologies for Environmental and Trade Reviews, OECD/GD(94)103*, Organization for Economic Co-operation and Development, Paris, France, 1994.
7. World_Bank, *The World Bank and Trade*, World Bank, Washington, D.C., 1995.
8. Daly, H. and Goodland, R., An ecological-economic assessment of deregulation of international commerce under GATT, *Ecol. Econ.*, 9(1), 3, 1994.
9. OECD, *The Environmental Effects of Trade*, Organization for Economic Co-operation and Development, Paris, France, 1994.
10. Beghin, J., Roland-Holst, D., and van_der_Mensbrugghe, D., *A Survey of the Trade and Environment Nexus: Global Dimensions*, OECD Development Centre, Organization for Economic Co-operation and Development, Paris, France, 1994.
11. Norgaard, R.B., Economics as mechanics and the demise of ecological diversity, *Ecol. Model.*, 38, 107, 1987.
12. Johnstone, N., Trade liberalization, economic specialization, and the environment, *Eco. Econ.*, 14(3), 161, 1995.
13. Copeland, B.R., International trade and the environment: policy reform in a polluted small economy, *J. of Environ. Econ. Manage.*, 26, 44, 1994.
14. Beghin, J., Roland-Holst, D., and van_der_Mensbrugghe, D., *Trade and Pollution Linkages: Piecemeal Reform and Optimal Intervention OECD/GD(94)90*, Organization for Economic Co-operation and Development, Paris, France, 1994.
15. World_Bank, *World Development Report*, Oxford University Press, New York, 1992.

16. Reed, D., Ed. *Structural Adjustment, the Environment, and Sustainable Development*, Island Books, Washington, D.C., 1996.

17. Browder, J., Ed., *Fragile Lands of Latin America: Strategies for Sustainable Development*, Westview Press, Boulder, Colorado, 1989.

18. Leonard, H.J., Overview, in *Environment and the Poor: Development Strategies for a Common Agenda*, H.J. Leonard, Ed., Overseas Development Council, New Brunswick, NJ, 1989.

19. Cruz, W. and Repetto, R., *Environmental Effects of Stabilization and Structural Adjustment Programs: The Philippines Case*, World Resources Institute, Washington, D.C., 1992.

20. Reed, D., Ed., *Structural Adjustment and the Environment*, Earthscan Books (for) the World Wide Fund for Nature, London, England, 1992.

21. OECD, *Joint Report on Trade and the Environment COM/ENV/EC/TD(91)*, Organization for Economic Co-operation and Development, Paris, France, 1991.

22. Abaza, H.J., UNEP/World Bank workshop on the environmental impacts of structural adjustment programmes, *Ecol. Econ.*, 14(1), 1, 1995.

23. Dessus, S., Roland-Holst, D., and van_der_Mensbrugghe, D., *Input-Based Pollution Estimates for Environmental Assessment in Developing Countries*, Organization for Economic Co-operation and Development, Paris, 1994.

24. Housman, R., *Reconciling Trade and the Environment: Lessons from the North American Free Trade Agreement*, Environment and Trade Series Paper No. 3, United Nations Environment Programme, Geneva, Switerland, 1994.

25. USTR, *Review of U.S.-Mexico Environmental Issues*, U. S. Trade Representative, Washington, D.C., 1992.

26. Harwell, C., et al., *Free Trade and the Environment: A Prospective Analysis and Case Study of Venezuela*, The North-South Center, Miami, Flordia, 1994.

27. CEC, *A Survey of Recent Attempts to Model Environmental Effects of Trade: An Overview and Selected Sources*, Prospectus Press (for) the Commission for Environmental Cooperation, Montreal, Canada, 1996.

28. Charnovitz, S., Dolphins and tuna: an analysis of the second GATT panel report, *Environ. Law Rev.*, 24, 10567, 1994.

29. Canada, *North American Free Trade Agreement: Canadian Environmental Review*, Canada, Ottawa, Canada, 1992.

30. USTR, *The GATT Uruguay Round Agreements: Report on Environmental Issues*, U.S. Trade Representative, Washington, D.C., 1994.

31. Charnovitz, S., NAFTA: An analysis of its environmental provisions, *Environ. Law Rep.*, 23, 10067, 1993.

32. USTR, *The NAFTA: Expanding U.S. Exports, Jobs, and Growth: Report on Environmental Issues*, U.S. Trade Representative, Washington, D.C., 1993.

8 Criteria for Evaluation of SEA

Wil A.H. Thissen

CONTENTS

INTRODUCTION

The interest in SEA, that is, EA at the level of proposed policies and/or plans, has grown significantly over the last several years. As its relevance is recognized more broadly, and experiences in practice accumulate, the need for systematic comparative and evaluative research increases. Core questions that may be answered by such research include: What can we learn from successes and failures in the past? What are the characteristics of success? What conditions are favorable to success? What methods work better than others? When? and Why? Recent overview publications on SEA have provided a basis by defining the main characteristics of the field and identifying elements of best practice.[1,2] Further empirical research is needed, as it may contribute significantly to the deepening and sharpening of the core body of SEA.

This is not as easy and straightforward as it may sound. Comparative research is complicated by the existence of a host of uncontrollable factors that make each situation different. Even the definition of criteria for a successful SEA is subject to debate, let alone the factors that determine the eventual success of an SEA effort.

This chapter will explore possible criteria for evaluation of SEAs, based on current literature on Environmental Impact Assessment (EIA) and SEA evaluation, extending the search to the field of policy analysis. While not exclusively or primarily focusing on environmental impacts, policy analysis has a number of characteristics in common with SEA: it aims to support decision-makers in strategic and policy contexts by providing policy-relevant information.

1-56670-360-3/00/$0.00+$.50

We will, first, briefly explore the differences and similarities between EIA, SEA, and policy analysis. Next, we will discuss the state-of-the-art with respect to evaluation and evaluation frameworks in EIA, SEA, and policy analysis. Finally, a number of observations will be made and conclusions drawn with respect to evaluation of SEAs. We will use the term evaluation in a broad sense, that is, covering both the intrinsic properties or qualities of an impact assessment as well as its effectiveness with respect to influencing the decision or policy process.

EIA, SEA, AND POLICY ANALYSIS

EIA and SEA both originated from a need to foster the attention for and weight given to environmental considerations in decision-making. EIA has been in existence for a few decades now, and in many countries an EIA procedure is required by formal regulations when projects of particular types and sizes are proposed that may have serious detrimental effects on the environment.

It has, however, been felt that EIA has its shortcomings as a tool to guarantee sufficient attention to environmental concerns in planning and policy-making. Applied at the project level, an EIA can usually affect only decision-making in a reactive way, attempting to block or modify a project proposal if it is environmentally harmful. An EIA process can, in general, not lead to entirely different proposals. SEA came into being as a reaction to these perceived shortcomings of EIA. Its purpose is to more structurally introduce and safeguard the attention to environmental concerns in the policy-making and planning phases of decision-making.[1,2]

Policy analysis originated in the 1950s in the U.S., in response to the need for a more rational basis to (public) policy decisions. The field has developed along different angles, as can still been noted from a number of basic texts on policy analysis: systems analysis and operations research,[3-7] economics and decision analysis,[8,9] and the social and political sciences.[10,11] The term "policy analysis" is at present used to indicate a wide spectrum of analytic or assessment activities, which have in common:[12]

- their objective is to support clients, often policy- or decision-makers
- their emphasis is on the collection, interpretation, and communication of information of relevance to a policy issue
- they are decision- or action-oriented, rather than science-oriented.

While these characteristics are quite similar to those of EIA and SEA, the key difference is in their underlying objectives. EIA and SEA have been developed specifically to enhance the attention and weight given to environmental concerns in decision-making. The objectives of a policy analysis may vary according to the type of analysis. Most so-called classic policy analytic studies intend to support decision-making from an independent perspective. For example, they do not put one interest or objective at the forefront, but rather intend to enlighten the decision process by providing a balanced and coherent perspective on alternatives, including their social, economic, environmental, and other impacts of relevance. But many policy analysis activities in practice are advocative, i.e., they provide information and arguments from the perspective of a particular actor or interest.[13] Like EIA and SEA, policy analysis as a field is still evolving. Recent literature emphasizes participation, stakeholder-oriented analyses, and the relevance of inclusion of the normative aspects and arguments in the analysis of policy debates, in addition to the more traditional science-based modes of analysis.[14]

While, in principle, policy analytic methods and approaches can be applied at both the project and the strategic or policy levels, most of the literature in policy analysis has been focusing on strategic and policy issues. Therefore, experiences and views on evaluation from the policy analysis field may be particularly relevant to SEA.

EVALUATION FRAMEWORKS AND STUDIES FOR EIA AND SEA

Various studies have recently been published relating to evaluation of the state of the art in EIA, ranging from evaluation of legal or institutional arrangements to evaluation of individual assessment processes and reports. We base our discussion on a limited selection of recent studies to provide an impression of the variety in approaches and criteria proposed and used for evaluation of environmental assessments. The primary focus is on identifying different types of evaluation criteria, i.e., aspects taken into account to develop a judgement on the goodness of the assessment. In addition, we will also pay attention to differences in the type of evaluation study and in the methods used to collect evaluative information.

We identified at least three types of evaluation studies. The first was studies evaluating and comparing a limited number of assessment studies in depth. The second was empirical evaluation studies based on a large sample of cases. The third was broad across-the-board "state-of-the-art" studies.

Examples of the first type include a study of 13 Environmental Statements (ES) for clinical waste incineration proposals in the U.K.,[15] and a cross-sectional study of ten EIA reports in Canada.[16] Both evaluations concentrate primarily on the contents of the assessment reports.

In a discussion of the quality concepts and aspects used, Lawrence[16] proposes an explicit distinction between "quality" and "effectiveness." Quality is used to assess the goodness of institutional arrangements, methods, and other "inputs," while effectiveness is concerned with the consequences ("outputs") of these. For evaluation of the quality of individual assessments, a distinction is proposed between processes, methods, and documents, the documents being the written products of the assessment activity. A list of quality criteria is developed, covering in particular, the way in which analysis and synthesis, evaluation, and impact management have been addressed in the study or report.

Effectiveness is related to direct and indirect outcomes. Direct outcomes refer to achievement of identified goals, actual realization of impacts and impact management measures as forecast, quality of proposals emerging from the process, compliance with regulations and commitments, and maintenance of environmental quality. Indirect outcomes may be assessed in terms of contributions to environmental management principles, administrative structures and cultures, research and science in a more general sense, and to the state of the art in EIA practice. The reader is referred to the original paper[16] for a detailed elaboration.

The evaluation is performed through application (by the author) of the criteria to the report contents, with several deficiencies and areas for improvement pointed out.

In addition to the contents of the ES, the U.K. study takes the presentation of the report, the kind of firm that performed the ES, and prior experience with ES of both the firm and the decision-making and planning organization into account.

Three different methods of evaluation were applied:

(1) an "objective" evaluation of the ES contents with reference to a basic checklist featuring aspects such as the number of alternatives considered, inclusion of baseline data, degree of quantification of impacts, assessment of impact significance, risk assessment, type and number of alternatives considered, and provisions for monitoring
(2) subjective assessments of ES quality by the analysts/authors of the assessment themselves
(3) subjective assessments of ES quality by planners associated with local planning authorities who are the intended users of the ES.

Of interest are the reported differences among the two groups in opinions about the quality of the assessments, as well as differences between the subjective assessments and the "objective" assessment results.

An example of the second type of evaluation study is provided by an empirical study of the impacts of 100 EIAs in the Netherlands.[17] The study makes a distinction between impacts of the assessment at the level of actions or actual behavior of actors, and impacts at the level of concepts or ideas. The first category concerns (adaptations in) visible and tangible actions resulting from or affected by the EIA process, while the second refers to impacts on views on the problem situation, reasoned arguments, etc. This study also makes a distinction between direct and indirect impacts. Direct impacts affect the actors involved in the decision-making process regarding the proposed activity (i.e., their actions, views, arguments). Indirect impacts affect situations other than those for which the EIA was originally carried out, such as general learning or knowledge acquisition, which may affect other processes in which an actor is involved.

A structured phone survey was at the core of the research methodology. For each EIA, different actors were interviewed, including the initiator, the competent authority, the secretary of the Environmental Impact Assessment Commission working group, and other actors such as consultants who had been involved. This resulted in a total of more than 600 respondents. Results were analyzed statistically to explore the existence of significant relations between the perception of impacts on the one hand, and (enabling) conditions on the other. The conditions explored were the formal characteristics of the process (division of roles, private/public initiator), the way the EIA was managed (starting moment, way of project initiation, communication), the initiative's degree of controversiality, and the existence of prior experience with EIA.

The study concludes that two EIA management conditions favor the effectiveness of EIAs: early initiation of the EIA, and the attunement of the efforts to the main policy themes rather than excessive detail. All other conditions do not appear to be related in a statistically significant way to the degree of effectiveness, at least in this sample.

A third, much broader type of approach has been taken in the International Study of the Effectiveness of Environmental Assessment.[18] Key points in the study include the notion that evaluations may address different levels:

- system-wide evaluations, addressing the adequacy of regulations, institutions, and general knowledge availability and dissemination
- evaluations of specific, single EA processes
- evaluations directed at specific aspects and components of EA studies, e.g., procedural compliance, documentation, methods used

The latter may be considered at the system-wide or the study-specific level. The evaluation framework adopted is elaborated in different ways. For example, a distinction is made between four perspectives on EA effectiveness:

- application of principles and provisions
- contributions to decision-making
- implementation of terms and provisions set out in the analysis
- eventual benefits to the environment

The institutional framework, available knowledge, operational capability, etc., are seen as enabling conditions for effectiveness for all four perspectives.

With respect to possible benefits for EA, a distinction is made between direct benefits and indirect benefits. Direct benefits, according to the study report, relate to the specific objectives of performing EAs: withdrawal of environmentally harmful proposals, modification and change of proposals, impact reduction, and environmental protection. Indirect benefits refer to non-intended but desirable effects such as community development, policy redefinition, and raising of environmental standards, or stimulation of research and development of environmentally appropriate technology.

The methodology for the study consisted of two parts: a number of surveys, and a series of workshops. The international survey covered EA practitioners, researchers and specialists, country status reports, and a survey of corporate use of EA.

The study conclusions and recommendations are extensive and refer to the variety of levels and aspects considered.

In a companion study specifically concentrating on SEA[1] the approach is equally broad, and the evaluation framework similar to that of the EA effectiveness study. A three-aspect approach is followed to evaluating SEA:

- adequacy of procedures, requirements, arrangements
- operational excellence referring to the rigor of the analysis, the quality and responsiveness of consultations, and the responsiveness or receptiveness of administrators/decision-makers
- relevance (does SEA inform choice, influence actions?)

Compared to the EA effectiveness study, relatively more attention is paid to the integration or matching of SEA with the decision-making process and culture. The difficulty of assessing the contributions of SEA is stressed. More than for EIA, where rules and provisions for results and handling have been specified, SEA influences are molded in a complex process where many other inputs play a role, and SEA contributions may be interpreted and valued differently by different actors.

In conclusion, there appears to be general agreement in the EA community that a distinction needs to be made between evaluations at the institutional or system level, and evaluations at the individual project or assessment level. The system level concerns regulations, institutions, habits, knowledge, and practices regarding EA in general. It provides conditions that affect the assessments and their impacts at the individual project level. Another field of agreement seems to be the distinction between what is often called "quality" at the level of implementation of regulations, application of methods, information products resulting from the EA, and the like, and "effectiveness" referring to the degree of influence of EAs on decision-making, and, ultimately, on environmental quality.[19]

Several authors make a distinction between "direct" and "indirect" effects, but they do not use these terms in identical ways. Some[17] make the distinction with respect to the actors affected and the subjects concerned, while others[16,19] use the terms to explicitly distinguish intended from unintentional but relevant impacts.

There appears to be remarkable differences in terms of the objects of evaluation (the EA process, the contents of the EA report, the impacts of the process, or a combination of these) and the number and type of perceptions from which the evaluation is made (a more or less objective evaluation only, or a larger variety of actors).

QUALITY CRITERIA IN THE FIELD OF POLICY ANALYSIS

Similar to what is seen in EA, a unified framework for evaluation of policy analysis efforts has not been established so far. The policy sciences offer a fundamental explanation: different views on the objectives of analysis and associated proposed evaluation criteria are based on fundamentally different views on what constitutes good policy-making and good support of policy-making.[12] Roughly, two main paradigms may be distinguished:

- policy analysis as a provision of objective information ("traditional policy analysis")
- policy analysis as a support of interactive negotiation processes in actor networks ("policy network approach").

In the traditional view, analysis provides (or should provide) independent and objective information. The underlying hypothesis is that policy-making is an activity in which an identifiable unicentral decision-maker or decision-making group makes one-time big decisions. The normative view is that these policy-makers should act rationally in terms of choosing the best means to the ends chosen by them. Hence, policies will get better as the available information on problems, causes, alternatives, and impacts is better, more scientific, and more balanced. The core objective of analysis is to generate and present such information. Analysis is primarily a cognitive analytic effort, which is based on objective scientific evidence and needs to be as independent as possible from the policy process. Choosing the ends and making (subjective) tradeoffs are not the responsibility of analysts, but of policy-makers.

We will discuss the key elements regarding evaluation and quality control from the traditional policy analysis perspective on the basis of two core publications on the subject.

First, Clark and Majone[20] make a distinction between three classes of evaluation criteria related to inputs, process, and outputs. Inputs are factors like resource and time constraints, data and theory availability, given problem formulations, quality of analysts, and the like. Process characteristics include methodology adopted, degree of formalization of analytic activities, knowledge utilization, attention to documentation, adherence to procedural rules, collaboration with users, and the like. Output characteristics for evaluation include, among others, quality and robustness of conclusions, clarity and accessibility of reports, matching with the issues and needs of the users, and the use of study results.

Second, Goeller[21] points out that different parties will have different goals and perspectives and will therefore evaluate the same analysis differently. Major parties at interest distinguished are the analysts, analytic peer groups, sponsors or clients, policy- or decision-makers, and public interest groups.

Goeller introduces the term "success" to indicate the goodness of analytic activities as well as their outputs and impacts, and distinguishes three different kinds of success:

- analytic success, which considers how the activity was performed and presented
- utilization success, which considers how the (results of the) activity were used in policy-making or implementation
- outcome success, which considers what happened to the problem situation as a consequence of the activity

Within each of these categories, various criteria and approaches may be used to evaluate the policy analytic activity. Analytic success may be determined by formal quality control and approval of selected parties. This may include components such as technical validity, persuasive validity, availability, credibility, timeliness, pertinence, and usefulness. The approval of selected parties is innately subjective. Peer reviewers will typically focus on scientific credibility, consistency, and the like, while stakeholders in the policy issue will judge the analysis from the perspective of adequate representation of their interest and usefulness to support their viewpoints. Utilization success concerns whether and how results were used, and for what purpose. Outcome success considers what happened to the problem situation and those affected by it as a result of this use. In particular, it considers the degree of improvement of the problem situation as a result of the implemented alternative, if the selection of that alternative was significantly influenced by the analysis.

Similar to a number of authors in EA, Goeller makes an additional distinction between direct success and indirect success. Direct success relates to impacts on the decision or problem for which the analysis was commissioned, while indirect success is related to other decisions and situations.

The policy network approach starts with the observation that the idea of a unitary decision-maker acting rationally is not in accordance with reality.

Policy formation and policy implementation are "inevitably the result of interactions among a plurality of separate, but mutually interdependent actors with separate interests, goals, and strategies."[22] It is even questionable as to what extent full rationality in the traditional sense would be desirable at all. It may be argued that not a single decision-maker or unitary decision-making group controlling (or attempting to control) society should make policies, but that the key stakeholders, themselves, should in mutual interaction strive for acceptable and implementable solutions. Depending on the dominant culture of the process, it may take the form of an open, participative, learning-oriented activity, or of a negotiation battle in which actors behave strategically in their attempt to achieve as much of their interests as possible.

In the case of an open, constructive process, the aim is for shared views on problems, objectives, and preferred solutions. Policies, in this view, get better if the stakeholder interaction process is more open and aimed at consensus reaching. The role of policy analysis, then, is to design and support an interactive multi-stakeholder process in which participants learn about the content of the issues as well as about each other's perceptions and needs. Substance-oriented arguments cannot be separated from the debate about values and the negotiation for interests, as these are closely intertwined.[14,23] As a result, evaluation criteria are proposed with respect to both process and contents of policy analysis activities.[24] Important content criteria are the variety of stakeholder perspectives and aspects taken into account, the use of state-of-the-art scientific knowledge, the consistency of arguments, the accessibility and understandability of outputs to a broad, non-academic audience, and similar analytic criteria stressing the broadness and openness of the discourse. Important process criteria proposed include:

- the involvement of relevant stakeholders in all phases of a policy analytic activity
- inclusion of a pluriform set of participants with different ideas and backgrounds
- the use of methods that stimulate open communication and allow participants to learn step by step
- the fit of the method of working with the rules and structure of the organization or policy network

With respect to the outcome, it is stressed that the selection or choice of a solution should be made within the context of all interests. Some authors writing about evaluation of interactive learning processes stress the necessity to evaluate these activities from the perspective of a variety of actors involved.[25] Others point out that learning effects may be distinguished at different levels, for example, the individual, the group that has participated, or the organization as a whole that may implement the conclusions.[26]

The idea that policy-making can be seen as a constructive and consensual learning effort is, however, often not realistic. Actors have their own problem perceptions, goals, and strategies, and try to steer toward their own preferences. Actors will often behave strategically and will therefore not engage in the kind of open and constructive dialogue presumed or aimed for by the interactive learning ideal. Then, evaluation criteria for support of policy development cannot be related to factors such as the degree of attainment of a commonly defined goal. Rather, process-related criteria are proposed, such as legitimacy, democratic nature, transparency and openness of the process, process continuity and efficiency. The costs of interaction and the level of commitment to the joint undertaking of those involved are also suggested as criteria.[27] Cognitive, research-based inputs may nevertheless be useful. Such analysis may affect the perceptions of individual actors, help to identify win-win options, contribute to the development of shared strategies and action plans, and contribute to breaking deadlocks in the interaction process by providing new viewpoints or options that may result in more satisfactory or acceptable outcomes for the parties participating in the network.[28,29] Other relevant effects of interactive processes may relate to changes in the structure of the network itself. New actors may have been introduced in a network, or changes may have developed in the relations between participants.

Goal attainment may be a measure of success or failure at the individual actor level (that is, actors achieving their own objectives). More generally, the extent to which participants consider the interaction and its results as satisfactory can be used as a criterion.[27] Ex-post satisfaction is related to the well-known criterion of achieving win-win situations, which is sometimes also proposed for judging the outcomes of negotiations in networks.

To summarize, like in EA, evaluation criteria of a policy analytic activity can and are being defined as attributes of the activity itself, of the direct results of the activity, of the way these results are being used, and of the eventual impact of the analysis on decision-making and problem resolution. Authors differ with respect to the terminology used, even the term "quality" is defined and explained in different ways. However, in general, the following conceptual structure for categorization of evaluation criteria at the level of individual policy analytic activities can be derived:[12]

- Input criteria relate to input conditions, i.e., aspects preceding or affecting the analysis, (e.g., who initiated the analysis, why the analysis was initiated, availability of data, institutional frameworks, etc.)
- content criteria relate to the content of the analysis, e.g., the validity of the analysis methods used, and the variety and relevance of alternatives and objectives that were considered in the analysis
- process criteria relate to characteristics of the (organization of the) analysis process, e.g., the transparency of the organization, the use of resources,

time, and money, and the cooperation among or involvement of various parties

- results criteria relate to the products of the analysis, e.g., the findings of the analysis, including the presentation, relevance, and validity of these results of the analysis
- use criteria relate to who uses which elements of the analysis for what purpose
- effects criteria relate to the possible effects of the policy analytic activity, e.g., whether the analysis fed the discussion, whether the analysis had any effect on the policy process, policy formulation, decisions taken, insights and arguments held by the players, improvement of the problem situation, or on other factors.

The underlying logic is pictured in Figure 8.1. A list of aggregate criteria for each of these categories is shown Table 8.1.

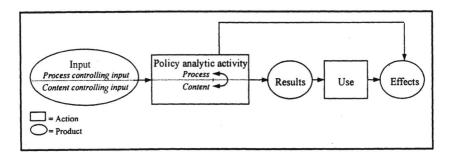

FIGURE 8.1   Framework for structuring evaluation criteria for policy analytic activities.

In their overview paper, Thissen and Twaalfhoven further point out that:

- Value judgements and associated quality criteria depend on context, time, and actor-perspective, or, "Quality is in the eye of the beholder." Analysts and scientists, for example, will often positively value a detailed and scientifically innovative impact analysis. Policy-makers and process managers may, however, not at all appreciate the same analysis as its outcomes may lead to further polarization, delays, or even deadlocks in the policy process.
- While most authors do agree that, after all, the ultimate impacts of a policy analytic activity on the policy process and its outcome are critical to the final evaluation, many concentrate on characteristics of the policy analytic activity itself when elaborating evaluation criteria. This is a more practical: it takes much more time and patience to wait for the ultimate effects. Moreover, as time elapses, it becomes more and more difficult to isolate the effects of the policy analytic activity amidst the variety of other factors affecting the policy process and hence its outcomes.

TABLE 8.1
Overview of Criteria Used for Evaluation of Policy Analytic Activities

Input

- Participants (type and quality of)
- Formal context of activity
- Institutional conditions
- Willingness/availability of actors
- Availability of time
- Availability of funds
- Availability of supporting tools
- Experience/quality of analysts
- Availability of data and knowledge

Content

- Adequacy of methodology
- Depth
- Broadness
- Validity, credibility
- Quality of argumentation
- Relevance

Process

- Parties involved
- Extent of cooperation
- Organization (flexibility, clarity)
- Working methods
- Efficiency, productivity
- Qualities of communication
- Compliance with regulations, formal procedures

Results or Products

- Consistency (internal)
- Relevance
- Presentation
- Availability
- Acceptance by parties involved

Use

- Which elements
- By who
- For what purpose

Effects

- Decision/policy content
- Policy process effectiveness
- Implementation of policies
- Benefits to problem situation
- Individual parties' ideas, arguments, insight
- Individual parties' well-being
- Collective insights
- Shared strategies, commitment
- Changes in social structure, network

Source: From[12]

But the substitution of effect-related criteria by criteria relating to the policy analytic activity itself and its inputs is risky. Analysis that is high quality from a scientific point of view does not necessarily lead to the desired or expected effects on policies or decisions, whereas efforts of an analytically less advanced nature may have considerable societal effects.

There is also a more fundamental reason for including inputs and process aspects in the evaluation. The inputs or process characteristics themselves may have a certain value to some parties, or may be objectionable to others.

- It is not feasible to aim for a single set of criteria to be used widely and unequivocally to evaluate all policy analytic activities. Rather, a subset of the rich variety of possible criteria may be used in individual cases, the

selection depending on context, time, actor-perspective, nature, and intentions of the specific activity.
- A structured overview of possible criteria, however, provides a helping
  hand to select the subset to be focused on. Designers of policy analytic
  activities may use such a list as a starting point to determine the specific
  characteristics and objectives the activity should aim for.

LESSONS FOR EVALUATION OF SEAS

As noted before, the key difference between SEA and policy analysis is in the
underlying objectives. EA has been developed specifically to enhance the attention
and weight given to environmental concerns in decision-making. Policy analysis
takes a broader approach, not putting one aspect centrally, but attempting to provide
a balanced and coherent perspective on alternatives, including their social, economic,
environmental, and other impacts of relevance. Despite this difference in objective,
many of the lessons and observations from the field of policy analysis do apply to
the evaluation of SEA, at the individual assessment level. The following points stand
out as possible contributions from policy analysis to SEA evaluation frameworks:

- The distinction between criteria at the level of inputs, the analytic activity
  itself, its direct results, and its ultimate effects (as illustrated in Figure
  8.1), and the overall structure and list of possible evaluation criteria shown
  in Table 8.1 are applicable to SEA evaluation as well.
- Many evaluation studies of EA concentrate on aspects such as analytic
  qualities and adherence to the rules rather than on the ultimate impacts
  of the assessment on decision-making. This entails the risk that factors
  of crucial importance to achieving those impacts are missed, and that
  important lessons may remain unlearned.
- Most evaluations–with a few exceptions–do recognize that different (types
  of) actors may use different criteria for evaluating EA. Nevertheless, they
  concentrate on broad brush evaluations in which the views of a heteroge-
  neous group of actors are aggregated somehow, or they focus on "objec-
  tivated" (i.e., analytic) criteria. The EA Evaluation Study, for example,
  based its conclusions largely on the opinions of two groups, namely those
  institutionally involved in EA, and those practicing or developing the craft
  of EA. This, inevitably, entails a supplier bias in criteria used as well as
  values given, and may not be beneficial to learning key long-term lessons
  about what decision-makers and other societal stakeholders appreciate
  about EA, and what not.
- While employing a broad set of criteria, few studies evaluating EAs
  explicitly pay attention to (non-intended) side-effects, some of which may
  be considered negative (costs, time, etc.), and some generally valued
  positively (improved communication, shared knowledge, changes in actor
  networks, etc.).
- Most (S)EA studies and their evaluations seem to fit best with the so-
  called traditional policy analysis paradigm, i.e., it is implicitly or explicitly

assumed that the provision of more and better science-based knowledge of impacts of alternatives will be instrumental in improving decision-making. Evaluation criteria focus on content and analytic rigor and less on process characteristics and efficiency. The underlying assumptions, however, will often not be valid, and lessons about supporting interactive negotiation processes may be very relevant, in particular to SEA. There is, indeed, a tendency in SEA to broaden the approach to include participation of citizens and other stakeholders. But additional emphasis on analysis of actor positions and identification of win-win situations may be fruitful.

- Those emphasizing the need for (public) participation clearly aim for improved communication and support for the assessment and decision afterwards. This is the kind of effect indicated as "changes in the social structure" in Table 8.1. In EA evaluation studies, this kind of effect has been generally ignored so far.

CONCLUDING REMARKS

Evaluation of EA activities is a complex endeavor. While there is general agreement that different levels of evaluation need to be distinguished, notably the institutional level and the individual assessment level, there is no generally agreed-upon framework for the evaluation of SEAs.

Similar difficulties are encountered in the field of policy analysis. In policy analysis, it is noted that different views on what is good policy-making result in different views on the role and contributions of analytic activities. As a result, different authors propose different sets of evaluation criteria. The variety of criteria can be structured in a general framework, making a distinction between inputs, process, content, results, use of results, and (ultimate) effects of the activity. This framework and the elaboration as shown in Table 8.1 are useful for SEA evaluations as well. Particular lessons for SEA evaluation include the relevance of differentiation between actor perspectives, the relevance of taking process criteria into account, and the need for an increased focus on the ultimate effects of the SEA rather than on the SEA process and analytic content alone.

ACKNOWLEDGMENT

The author thanks Patricia Twaalfhoven and the editors for useful comments on an earlier draft of this chapter.

REFERENCES

1. Sadler, B. and Verheem, R., *Strategic Environmental Assessment. Status, Challenges and Future Directions*, Ministry of Housing, Spatial Planning and the Environment, 53, Zoetermeer, the Netherlands, 1996.

2. Therivel, R. and Partidário, M.R., *The Practice of Strategic Environmental Assessment*, Earth Scan Publications, London, 1996.
3. Quade, E.S., *Analysis for Public Decisions*, 3rd ed., North Holland, 1989.
4. Miser, H.J. and Quade, E., *Handbook of Systems Analysis. I. Overview of Uses, Procedures, Applications, and Practice*, North Holland, 1985.
5. Miser, H.J. and Quade, E.S., Toward quality control, in *Handbook of Systems Analysis, Craft Issues and Procedural Choices*, Miser, H.J., and Quade, E.S., Eds., John Wiley & Sons, 619–656, 1998.
6. Miser, H.J. and Quade, E., *Handbook of Systems Analysis. II. Craft Issues and Procedural Choices*, Wiley, 1988.
7. Miser, H.J., Ed., *Handbook of Systems Analysis. III Cases*, Wiley, Chichester, 1995.
8. Patton, C. and Sawicki, D., *Basic Methods of Policy Analysis and Planning*, Prentice Hall, 1986.
9. Stokey, E. and Zeckhauser, R., *A Primer for Policy Analysis*, Norton, New York, 1978.
10. Dunn, W., *Public Policy Analysis. An Introduction*, Prentice Hall, 1981.
11. Wildawsky, A., *Speaking Truth to Power. The Art and Craft of Policy Analysis*, 2nd ed, Transaction Publishers, 1993.
12. Thissen, W. and Twaalfhoven, P., Towards a conceptual structure for evaluating policy analytic activities, *Eur. J. Oper. Res.*, submitted.
13. Thissen, W., From SEA to integrated assessment: a policy analysis perspective, *EA, Mag. of the Inst. of Environ. Assess. and the Environ. Aud. Regist. Assoc.*, Lincoln, U.K., 25–26, 1997.
14. Fischer, F. and Forester, J., Eds., *The Argumentative Turn in Policy Analysis and Planning*, Duke University Press, Durham, North Carolina, 1993.
15. Radcliffe, A. and Edwards-Jones, G., The quality of the environmental assessment process: a case study on clinical waste incineration in the U.K., *Proj. Apprais.*, 10(1), 31–38, 1995.
16. Lawrence, D.P., Quality and effectiveness of environmental impact assessments: lessons and insights from ten assessments in Canada, *Proj. Apprais.*, 12(4), 219–232, 1997.
17. ten Heuvelhof, E. and Nauta, C., The effects of environmental impact assessments in the Netherlands, *Proj. Apprais.*, 12(1), 25–30, 1997.
18. Sadler, B., *Environmental Assessment in a Changing World: Evaluating Practice to Improve Performance*, Canadian Environmental Agency/IAIA, Minister of Supply and Services Canada, cat. no.: EN106-37/1996E, 1996.
19. Sadler, B., *Ex-post Evaluation of the Effectiveness of Environmental Assessment*, Chapter 3 in A.L. Porter and J.J. Fitipaldi, Eds., *Environmental Methods Review: Retooling Impact Assessment for the New Century*, The Press Club, Fargo, North Dakota for IAIA and AEPI, 1998.
20. Clark, W.C. and Majone, G., The critical appraisal of scientific inquiries with policy implications, *Sci., Tech. Human Values*, 10, 6, 1985.
21. Goeller, B.F., *A Framework for Evaluating Success in Systems Analysis*, in *Handbook of Systems Analysis, Craft Issues and Procedural Choices*, Miser, H.J. and Quade, E.S., Eds., John Wiley & Sons, 567–618, 1988.
22. Sharpf, F.W., Reissert, B., and Schnabel, F., Policy effectiveness and conflict avoidance in intergovernmental policy formulation, in *Interorganizational Policy Making: Limits to Coordination and Cental Control*, Eden, C. and Akerman, F., Eds., Sage, London, 57–114, 1978.
23. White, L.G., Policy analysis as discourse, *J. of Policy Analy. and Manage.*, 13(3), 506–525, 1994.

24. Geurts, J.L.A. and Kasperkovitz, J.M., The science/public policy dialogue on long-term environmental planning, in Dutch Committee for Long Term Environmental Policy (CLTM), *The Environment: Towards a Sustainable Future*, Kluwer Academic Publications, 113-141, Dordrecht, 1994.
25. Eden, C. and Ackermann, F., Horses for courses: a stakeholder approach to the evlautaion of GDSSs, *Group Decision and Negotiation*, 5, 501, 1996.
26. Andersen, D.F., Richardson, G.P., and Vennix, J.A.M., Group model building: adding more science to the craft, *Syst. Dyn. Rev.*, 13(2), 187–201, 1997.
27. Kickert, W.J.M., Klijn, E.H., and Koppenjan, J.F.M., Eds., *Managing Complex Networks; Strategies for the Public Sector*, Sage Publications Ltd., London, 1997.
28. Klijn, E.H., Koppenjan, J.F.M., and Termeer, C.J.A.M., Managing networks in the public sector: a theoretical study of management strategies in policy networks, *Public Adm.*, 73(3), 437–54, 1995.
29. Jasanoff, S., *The Fifth Branch, Science Advisers as Policy Makers*, Harvard University Press, Boston, Massachusetts, 1990.

Section IV

Procedural Approaches to SEA at a Plan and Program Level

9 SEA Experience in Development Assistance Using the Environmental Overview

A. Lex Brown

CONTENTS

INTRODUCTION

This chapter describes experience with the Environmental Overview as an appropriate and highly effective form of Strategic Environmental Assessment (SEA) in the field of development assistance. The origins of the Environmental Overview, the niche that it occupies in the environmental assessment *family* of project and program appraisal tools, and details of the methodology of the Environmental Overview, have been described elsewhere.[1,2] In brief, the Environmental Overview is a particular form of SEA that was developed to appraise the environmental consequences of programs internally within the United Nations Development Program (UNDP). Here, the need was for a method of appraisal that could be applied to a very wide range of capacity-building and other, generally non-hardware, development plans and programs, and that could be deployed rapidly, effectively, at low cost, and on a regular basis. To achieve this, the Environmental Overview utilizes a small group approach where its participants examine the proposal and its context, scope the potential social and environmental consequences and opportunities associated with the proposal, and propose how it can be modified to reduce the unwanted consequences and to enhance the opportunities. The involvement of a broad cross-section of interests and disciplinary skills in the small group is fundamental to its operation.

While it was prepared and refined within a specific organizational context, the Environmental Overview now has widespread application as a tool for the appraisal of development plans, policies, and programs (PPP) well beyond its origins, particularly where there is need for a rapid or preliminary assessment. In many jurisdictions and in many circumstances, particularly in, but not restricted to, the developing world, unless rapid SEA techniques are available, many PPPs will move from formulation to decision-making without the application of *any* form of environmental appraisal. Of course, if the decision-making process is such that time and resources are procurable, more extended forms of SEA would be preferred, but it is vital, where this is not the case, that a formalized, well-structured, and theoretically sound, methodology of SEA is available. The Environmental Overview performs this role. In this chapter, consideration is given to why the Environmental Overview is an effective and efficient tool for SEA as part of the *formulation* of development PPPs.

WHY APPLY SEA TO DEVELOPMENT ASSISTANCE PROGRAMS?

While various countries are experimenting with different SEA approaches, experience is as yet too limited to conclude the effectiveness of such systems[4] or to specify the most desirable approach toward strategic assessments.[5] Today, there is rapid transfer of technology from the developed world to the developing world, and to countries in transition, and this transfer includes the "technology" of environmental assessment (EA) methods. With so much emphasis in the current literature on SEA, attempts to transfer this approach to program appraisal in the development context will be widespread. A new tool is always seductive both to practitioners and to administrators in developing countries, and western experts are not shy to offer a new product, even if that product is still little more than a broad notion in most developed countries. There is a danger that SEA could be adopted rapidly and unquestioningly in developing countries in the same wholesale fashion as was the project Environmental Impact Assessment (EIA)[5] originally, without trial and adaptation to local needs, and without overall consideration as to what will function institutionally in developing countries.

Project-based EIA is now applied almost universally, and it is most unlikely that major project developments supported by international or bilateral donors anywhere in the world would progress without EIA as part of project appraisal. But whereas aid in the past has been in the form of projects amenable to project-based EIAs, most aid today is "softer." It is not project-based. It may be programatic — a range of inter-related activities under a single theme. It may provide wide-ranging support to a sector or sub-sector of government. Moreover, it may focus primarily on in-country capacity building, both through human development and institution building, over the complete range of governmental and non-governmental activities. Projects, or programs, such as *developing the capacity to privatize state-run enterprises, technical assistance to develop a fishing industry, promotion of export growth,* or *capacity development for land-use planning,* are the norms. Development assistance of this type always falls outside of the requirements for project-based EIA, but it is not immune from potential long-term environmental and social externalities arising from its activities — just as physically based projects are not. It, too, needs appraisal. Soft aid deals with activities which occur upstream in the decision-making cycle. It deals with whole programs, even whole government sectors. Often, development assistance, though a capacity-building focus, is part of policy development. Clearly, these shifts in the nature of aid require assessment tools which operate at a more strategic level than project-based EIA. The question this chapter addresses is, "do we have any tools in the family of EA methods and experience that can be applied to these types of soft projects/programs in overseas development aid?"[5]

There is world-wide interest in this question. An Organization of Economic Co-Operation and Development (OECD) Working Party[6] has examined the role of SEA in overseas development assistance. In 1992, Lee and Walsh[7] noted that some of the multi-national and bi-national aid agencies and banks were showing interest in the extension of EA to the more strategic levels of planning and decision-making,[8,9] but actual practice to that time was fairly limited. Since then, the World Bank has had more experience through its major contributions in Sectoral Environmental Assessment[10] and Regional Environmental Assessment.[11] However, experience of SEA is still very limited in terms of its application across much of the range of development assistance activities. The Australian Agency for International Development (AusAID), in its most recent revision of its EA guidelines for development aid,[12] has included the term SEA as a step in its overall assessment program. This is intended to provide a higher level of environmental analysis of environmental PPP, but closer examination reveals that AusAID is also looking for methodologies to implement its intentions. Scanlon,[13] Mahony,[14] and Kennett and Perl,[15] all report the need for upstreaming EA in development assistance programming activities.

THE ENVIRONMENTAL OVERVIEW

The UNDP now has considerable experience in the EA of aid of this type. This has been through development of environmental management guidelines[16,17] and an associated extensive training program for field staff and national government counterparts. The core of the experience is the Environmental Overview.[1] This process is used in the formulation stages of programs, which leads to early identification of environmental and social impacts and opportunities of those programs, and the incorporation of steps to mitigate those impacts or enhance those opportunities directly into program redesign. It must be undertaken participatorily, using a broad mix of specialists and others. The process should include modification of the draft project/program as an integral part of the Environmental Overview. While this may appear to be an optional step, the author's experience is that the appraisal becomes far more effective and efficient when the participants in the activity move iteratively between critique of the draft proposal and its modification, mitigating the identified impacts and incorporating the identified opportunities. This step encourages true creativity. SEAs must be recognized as creative processes, not just as documents, conforming to Partidário's[3] view that the objective of SEA must be much more than the production of an SEA report. The *interactive process* within the Environmental Overview, including any consequential changes to the project/program, is the heart of the technique.

The Environmental Overview asks a set of questions, similar to those asked by conventional project EIA, but with different emphasis. Fundamentally, these questions align with those "key requirements for conducting a strategic environmental assessment" identified by the OECD,[6] though there are differences in emphasis, with the Environmental Overview requiring baseline conditions on current economic forces and current management capabilities, as well as more emphasis on implementing the interventions.

The essence of the Environmental Overview lies in the wealth of expertise that can be rapidly brought to bear on a proposed development by interactively involving a range of parties in a group, and by this group's constructive dynamics. Through training experience, the Environmental Overview has been found (Box 9.1) to have applicability to a very wide range of activities, in terms of scale, nature, and geographical setting.

BOX 9.1

Examples of Programs/Policies Subject to Appraisal by the Environmental Overview as Part of Training Activities

- tourism development and management (Cambodia & Tonga)
- planning for the resettlement of tsetse-fly cleared area (Zimbabwe)
- state enterprises reform program: privatization (Vietnam)
- institutional support on the implementation of the National Shelter Strategy (Indonesia & Namibia)
- essential oils project (Bhutan)
- improvement of land settlement schemes (Mekong Secretariat)
- achieving international competitiveness through technology transfer and development (Philippines)
- technical assistance to the Roads Branch (Swaziland)
- employment generation through development of small, medium, and micro enterprises (South Africa)
- regional development policies for a province (Thailand)
- a set of policies for handling urbanization issues (Africa)

The OECD[6] notes that SEA is subject to a wide degree of procedural interpretation when applied to development assistance activities, but this is reasonable, given the flexible interpretation of what constitutes an SEA even within most developed countries. The Environmental Overview represents an innovative form of SEA in the development context. It is not surprising that new forms of SEA must emerge to meet particular needs. Sadler[18] observes that the promotion of a single procedural model is an inappropriate response to the many circumstances and configurations of policy-making.

WHY THE ENVIRONMENTAL OVERVIEW WORKS AS SEA

There appears to be a convergence in views that EA at the strategic level cannot be simply an expansion of concepts, processes, and legislation as currently apply for project-level EIA. Instead, it must take on quite different modes of triggering, functioning, and outcomes, borrowing only fundamental principles, not form, from project-based EIA. The Environmental Overview borrows from project-based EIA in its systems view, its multidisciplinary approach, and its structured analysis. It differs from project-based EIA in its participatory, interactive approach, short time-frame, low cost, and de-emphasis of the final report.

Because SEA operates at a different point in the planning process, and at a different level of generality than does EIA, Lee and Walsh's[7] expectation that there will be procedural and methodological differences between them is reasonable. Therivel et al.[19] argued that there are few officially acknowledged methods for SEA, and new forms and procedures relevant to particular settings need to be urgently developed and tested. They also argued that simple methodological approaches are required.

It is most unlikely that any of the programs shown in Box 9.1 would have required, or would have been amenable to even if required, conventional project-based EIA, yet each needed to have its environmental and social consequences appraised. Why is it that the simple Environmental Overview approach to evaluation appeared to work in such a wide range of development contexts?

It could be that the participatory discussion required by the Environmental Overview on problems and possible solutions, fits more comfortably with the way conflicts are resolved in the cultures of many developing countries, than do the more analytical and aloof processes provided by EIA-type approaches. Beyond this, it can be argued that the Environmental Overview may work because it conforms to many of the emerging principles for effective SEA espoused by contemporary writers on the topic. These principles, summarized in Box 9.2, are described in more detail below.

BOX 9.2
Some Principles for Effective SEA

- should not delay strategic decision-making
- must conform to the continuous, iterative, and evolutionary nature of policy and program development and decision-making
- must not stand alone but must be integrated into policy-making/decision-making
- must focus on process not product
- should encourage examination of alternative PPP
- may not require prediction based on sophisticated modelling
- is more complex than project-based EIA, but the degree of detail and level of accuracy required is generally less
- requires a level of cross- and inter-departmental communication and cooperation with which we are not familiar
- must overcome the potentially threatening nature of SEA to many agencies as this will be one of the largest impediments to its effective implementation.

EA should not cause delay in strategic decision-making because of the time it might take to prepare a strategic EIA. In strategic decisions there is no single decision-making moment as there often is with project decisions. Policy is continually added to, modified, or even withdrawn. Therefore, a strategic EIA must be able to be drafted quickly to provide the right information at the right time in this continuous process. It must not need years of study to provide this information.[4] It has been noted too,[4] that the Dutch Etest, asking the right questions during policy formulation, is one way of doing this. Sadler[20] also notes that SEA must not stand alone from decision-making, but should directly integrate EA principles directly into the decision-making process. The focus of the SEA must be on process, not on product.[3] Rather than the production of an SEA report, effort must be on an iterative and continuous process that assists the on-going policy-making process. SEA is intended to look at a range of possible alternatives in PPP in a way that is systematic and ensures full integration of relevant issues in the total environment including biophysical, economic, social, and political considerations. SEA must be an aid to policy formulation, rather than a post-formulation approach to mitigation, and environmental analysis must therefore be as intrinsic an element of policy formulation and analysis as is economic analysis.[3]

Dixon and Montz[21] argue that there is a reasonable premise that prediction based on sophisticated modeling may not be required in strategic assessment to enhance decision-making. Lee and Walsh[7] point out that differences in scale increase the complexity of SEA relative to EIA, but the degree of detail and the level of accuracy of information needed for PPP decision-making is generally less than that needed for project evaluation.

Cuff and Ruddy[22] point out that effective SEA requires a level of cross- and inter-departmental communication and co-operation with which our present political administrators are not familiar. SEA does not fit easily into the current departmental structure of most governments, and its development will require mechanisms which bring disparate parts of government together to consider impacts and to involve them creatively in the formulation of programs and policies. Ortolano and Shepherd[23] note that such ideas are threatening to line agencies, and this could be one of the largest impediments to effective implementation of SEA. Proponent agencies will be wary of giving potential opponents too complete a perspective of program impacts

The Environmental Overview process meets each of the imperatives in Box 9.2 because of its flexibility, speed, emphasis on process and integration with policy/program formulation, reliance in the first instance on environmental information already known by the group participants, and creative encouragement of consideration of alternatives and policy/program modifications. In fact, it meets nearly all of Partidário's[3] list of policy framework, institutional, and procedural issues, which an effective SEA system would need to address. It also fits closely with SEA requirements described by Verheem.[4] Further, by judicious selection of the participatory group conducting the Environmental Overview, and it has already been pointed out that the involvement of a broad cross-section of interests and disciplinary skills in the small group is fundamental to its operation, the procedure can go a long way toward achieving the necessary inter-departmental communication for effective SEA.

The Environmental Overview assessment tool represents a trade-off between a coarse but effective, tool to apply frequently and to a wide range of assessments, and a more comprehensive tool such as project-based EIA which is applied infrequently and to a restricted range of activities. It has proved to be a rapid appraisal tool for development proposal formulation, and is a mechanism for directly incorporating impacts and opportunities illuminated by this process into redesign of the proposal.

The Environmental Overview is likely to prove applicable to capacity-building activities, structural adjustment programs, and feasibility studies for project-based developments. It also works on land-use planning policies and sectoral development problems such as urbanization. It works at any scale: whether at project level, program level, or country level. It is applicable to hardware projects,software projects such as capacity building, and even to the assessment of policies. Of course it is no panacea, and requires commitment to the process by the proponent, by environmental agencies and government as a whole. No development planning tool or methodology is immune from abuse in practice.[24]

The Environmental Overview has evolved from training experience in development activities, but may prove to be a very versatile tool. Some limited experience has shown that the Environmental Overview can be used outside of the UNDP system and for a wide range of policy, program, and sectoral analyses other than aid projects. It has the potential to be a model that governments themselves can adopt and adapt as appropriate to their own internal development planning procedures. It should be recognized as a highly appropriate form of SEA for use in developing countries. The Environmental Overview is at the cutting edge of SEA of PPP in a development context.

CONCLUSIONS

SEA is currently a much talked about, but little practiced, tool and this creates difficulties for developing countries in that there are strong pressures for the rapid and often indiscriminate transfer of new "technologies" such as SEA from the developed to the developing world. The *Environmental Overview* is a new environmental assessment tool that has been extensively trailed in training programs in developing countries for the assessment of development of assistance projects. It is a tool that has been home grown in the development context.

The Environmental Overview represents the development of a form of SEA suitable for the routine and rapid environmental appraisal of programs, sectors, or policies in an international development aid context — an area in which there has been little experience to date. It is a participatory creative process, used in the formulation stages of development activities, that leads to early identification of environmental and social impacts and opportunities of those programs and in direct feedback into program redesign.

Beyond application to development aid, and application by developing countries to their internal programs, the Environmental Overview may prove to be the base model on which to construct an important member of that family of elusive SEA tools currently being sought in developed countries for the environmental appraisal of PPPs. It fits closely to the sets of requirements put forward by many commentators for an effective approach to SEA.

REFERENCES

1. Brown, A.L., The Environmental Overview in development project formulation, *Impact Assess.,* 15, 73, 1997.
2. United Nations Industrial Development Organisation, Environmental considerations in project design: Learning unit 9, *UNIDO Training Course on Ecologically Sustainable Development,* 1994.
3. Partidário, M.R., Strategic Environmental Assessment: Key issues emerging from recent practice, *Environ. Impact Assess. Rev.,* 16, 31, 1996.
4. Verheem, R., Environmental assessment at the strategic level in the Netherlands, *Proj. Apprais.,* 7, 150, 1992.
5. Brown, A.L., Environmental Impact Assessment in a development context, *Environ. Impact Assess. Rev.,* 10, 135, 1990.
6. OECD, Strategic Environmental Assessment (SEA) in Development Co-operation: State-of-the-Art Review, Draft Final Report, OECD/DAC Working Party on Development Assistance and Environment, OECD, Paris, 1997.
7. Lee, N. and Walsh, F., Strategic Environmental Assessment: An overview. *Proj. Apprais.,* 7, 126, 1992.
8. World Bank, *Environ. Assessment Sourcebook Update,* Number 1, Environment Department, The World Bank, 1993.
9. Goodland, R. and Edmundson, V., *Environmental Assessment and Development,* A World Bank-IAIA Symposium, Washington, The World Bank, 1994.
10. World Bank, Sectoral Environmental Assessment, *Environmental Assessment Sourcebook Update,* Number 2, 8, 1993.
11. World Bank, Regional Environmental Assessment, *Environmental Assessment Sourcebook Update,* Number 15, 10, 1996.
12. AusAID (Australian Agency for International Development), *Environmental Assessment Guidelines for Australia's Aid Program,* Commonwealth of Australia, 1996.
13. Scanlon, J.E.L., AIDAB and EIA: From rhetoric to action, *Environmental and Planning Law Journal,* 11, 492, 1994.
14. Mahony, S., World Bank policies and practice in Environmental Impact Assessment, *Environ. and Plan. Law J.,* 12, 97, 1995.
15. Kennett, S. A. and Perl, A., Environmental Impact Assessment of development-oriented research, *Environ. Impact Assess. Rev.,* 15, 341, 1995.
16. United Nations Development Program, *Handbook and Guidelines for Environmental Management and Sustainable Development,* New York, Environment and Natural Resources Group, UNDP, 1992.
17. Development Program's Environmental Management and Sustainable Development in Technical Assistance (EMG): Four international experts respond, *Environ. Impact Assess. Rev.,* 12, 139, 1992.
18. Sadler, B., Environmental assessment and development policy-making. In *Environmental Assessment and Development,* Goodland, R. and Edmundson, V., Eds., An IAIA-World Bank Symposium, Washington, DC, The World Bank, 1994.
19. Therivel, R., Wilson, E., Thompson, S., Heaney, D., and Pritchard, D., *Strategic Environmental Assessment,* London, Earthscan, 1992.
20. Sadler, B., *International Study of the Effectiveness of Environmental Assessment, Environmental Assessment in a Changing World: Evaluating Practice to Improve Performance,* Final report, Canadian Environmental Agency & International Association for Impact Assessment, 248, 1996.

21. Dixon, D. and Montz, B., From concept to practice: implementing cumulative impact assessment in New Zealand, *Environ. Manage.,* 19, 445, 1995.

22. Cuff, J. and Ruddy, G., SEA – evaluating the policies EIA cannot reach, *Town and Ctry. Plan.,* 63, 45, 1994.

23. Ortolano, L. and Shepherd, A., Environmental Impact Assessment: challenges and opportunities, *Impact Assess.,* 3, 13, 1995.

24. Wiggins, S. and Shields, D., Clarifying the 'logical framework' as a tool for planning and managing development projects, *Proj. Apprais.,* 10, 1995.

10 SEA of Parks Canada Management Plans

Suzanne Therrien-Richards

CONTENTS

BACKGROUND

Parks Canada is responsible for Canada's national parks, national marine conservation areas, and national historic sites, including historic canals. There are currently 38 national parks and national park reserves in Canada, located in every province and territory ranging in size from 8.7 to 44,802 km², covering a total of 224,466 km² or about 2% of Canada's land mass. In addition, 792 historic sites have been designated as nationally significant. The system of National Marine Conservation Areas is relatively young, with only 5 of 29 marine regions represented by 4 sites to date, covering a total of 4,518 km².

Parks Canada's purpose is stated as follows in the Guiding Principles and Operational Policies of the Department of Canadian Heritage: "To fulfill national and international responsibilities in mandated areas of heritage recognition and conservation; and to commemorate, protect and present, both directly and indirectly, Canada's natural and cultural heritage in ways that encourage public understanding, appreciation and enjoyment of this heritage, while ensuring long-term ecological and commemorative integrity."[1]

1-56670-360-3/00/$0.00+$.50
© 2000 by CRC Press LLC

The sustainable use of the nation's protected heritage places is further identified as an objective within the sustainable development strategy for the Department of Canadian Heritage.[2] The commitment to protecting ecological integrity and ensuring commemorative integrity is one of the guiding principles, which is adhered to in all aspects of decision-making within Parks Canada, from the development of policy to the implementation of projects.

Strategic directions for the national parks, marine conservation areas, and national historic sites are established through management plans. The zoning system is an important management strategy within a management plan to protect lands and resources and ensure a minimum of human-induced change by establishing whether or not certain activities will be allowed within certain zones. The planning objectives established through the management planning process, combined with zoning decisions, will determine the sustainability of Parks Canada's heritage places for future generations. The strategic environmental assessment (SEA) of management plans, therefore, becomes an effective mechanism by which to ensure that the purpose of Parks Canada is not compromised by initiatives within its mandated areas, integrating environmental and sustainability factors early in the planning process.

Environmental Assessment Framework

The Parks Canada Management Directive, 2.4.2. Impact Assessment,[3] sets out the legislative and policy framework for conducting environmental assessment (EA) within Parks Canada. A non-legislated process entitled, The 1999 Cabinet Directive on the Environmental Assessment of Policy Plan and Program Proposals, established by the Government of Canada, governs the assessment of policies, plans, and programs that may result in environmental effects.[4] A second regime, which complements the Cabinet Directive, is the Canadian Environmental Assessment Act, proclaimed in 1995, which applies specifically to projects. In addition, various instruments or authorities including activity policies specify impact assessment requirements for projects within Parks Canada which do not fall within the scope of the Canadian Environmental Assessment Act. Collectively, these regimes ensure that impact assessment is fully integrated throughout all policy, program and project development, and decision-making in Canada's heritage places.

The Parks Canada Management Directive 2.4.2 provides further clarification respecting SEA, stating: "the impact of management plans, business plans and related planning products and policies will be assessed in conformance with the Government of Canada Cabinet Directive."[3] This direction is important since management plans within Parks Canada contain the management objectives to protect and represent the natural and cultural resources of nationally important areas.[1,5] Within national parks, management plans specify the type and degree of resource protection and management; set out the type, character, and location of visitor facilities, activities and services; and identify target groups. In addition, the management plan sets out the framework for further planning processes respecting ecosystem management and interpretation, visitor services, and visitor risk management. A management plan for a marine conservation area provides guidance to managers and users about the day-to-day management and use of the area, while conserving the area's resources.

Management plans for national historic parks, sites, and canals are intended to guide managers in the protection and presentation of the site, since protection and presentation are viewed as fundamental to the commemorative integrity of a site.

Management Planning Process

Management plans for national parks and marine conservation areas are tabled in Parliament within five years after the proclamation of a site, or within five years of the transfer of lands proposed for establishment as a national park, and are reviewed every five years thereafter. Management plans for national historic parks, sites, or canals are currently reviewed every ten years, although an amendment to this time frame is under consideration. However, when significant changes in federal policy or circumstances affect the implementation of any management plan, a review can be initiated. The review of a management plan allows for the monitoring of the implementation and effectiveness of the management plan in the sustainable use of nationally significant heritage resources.

One of the management strategies used within national parks and marine conservation areas in the context of management planning to maintain ecological integrity is the assignation of zones. This zoning system applies to all land and water areas of national parks, and to other natural areas within Parks Canada, including national marine conservation areas. It consists of an integrated approach to classify areas based on both their ecosystem and cultural resource protection requirements and their capacity to provide sustainable visitor experiences. The concept of zoning recognizes the need to apply the principle of ecological integrity while recognizing the importance of providing appropriate visitor activities to enhance public understanding, appreciation, and enjoyment of heritage areas with a minimum of human change. Zoning does not preclude traditional harvesting rights and activities that have been established by land claims agreements, national park reserve status, or new park establishment agreement. Rather, zoning reflects the range of activities that may occur in heritage places, from the need for wilderness areas receiving minimal human interference to the demand for areas with intense visitor activities and supporting services.

In national parks, the zoning system consists of five zones as follows:

- *Zone I:* Special Preservation is reserved for specific areas or features requiring preservation due to the presence of unique, threatened, or endangered natural or cultural features. This zone is also used to designate areas within a national park which are particularly representative of the natural region for which the park has been recognized. Motorized access is not allowed within Zone I.
- *Zone II:* Wilderness areas represent natural regions to be conserved in a wilderness state and together with Zone I constitute the majority of the area of most national parks. The key consideration is minimal human interference; motorized access and circulation are not allowed, with the exception of controlled air access in remote northern parks. Opportunities for visitors to experience the park are based on maintaining the wilderness value, with the provision of few rudimentary services and facilities.

- *Zone III*: Natural Environment areas are managed to provide opportunities to visitors through minimal services and facilities of a rustic nature, while allowing the visitor to experience the natural and cultural heritage. Motorized access is controlled, and public transit is encouraged.
- *Zone IV*: Outdoor Recreation is characterized by direct access by motorized vehicles to essential services and facilities for the public.
- *Zone V*: Park Services recognizes communities or town sites in existing national parks that provide a concentration of visitor services and facilities.

In addition to the five zoning designations, Parks Canada policy recognizes the need to designate "environmentally or culturally sensitive areas" to implement special management practices for areas or sites which do not fit the zoning designations above. The objective of the designation is to allow the prescription of management tools that will protect the sites' integrity as a significant natural or cultural resource and to present and interpret these resources in a way that respects the resource's unique values and needs. As an example, a management prescription could include the prohibition of visitor access on a seasonal basis, or the need to conduct intensified research and monitoring.

However, the zoning system as described above may not be appropriate for many national historic parks, sites, and canals for a number of reasons. Many of these sites will have been recognized and commemorated for reasons which are inconsistent with "wilderness" or "natural environment" objectives. The sites may be set in an urban environment, where such objectives are inappropriate. Alternately, a reduced land base may preclude the establishment of multiple use zones. A modified, simpler version of the zoning system has been developed, and it provides a useful mechanism for some sites to guide decision-making with respect to the appropriateness of proposed activities/uses. The site is organized into only two zones, a contemporary zone and a historic zone. The contemporary zone contains modern facilities such as visitor reception centers, parking lots, picnic areas, administrative centers, etc., whereas the historic zone contains extant historic structures, archaeological resources, cultural landscapes, and other cultural resources.

The value of conducting impact assessments of management plans quickly becomes apparent. Management plans provide the strategic direction for the sustainable use of nationally significant areas, attempting to balance economic opportunities for services, facilities, and access to the public while ensuring that the ecological and commemorative integrity are maintained for future generations of Canadians. The SEA of the management plans provides the mechanism to ensure that the heritage values of significant areas are protected from the pressures for economic development using such tools as the zoning system to provide direction for appropriate decision-making.

The EA of a management plan, therefore, represents the application of SEA as suggested by several authors whereby the concept of integrating socio-economic aims into environmental considerations is applied in policy-making, planning, and program development.[6-9] Partidário describes this approach as the top-down approach, where EA principles are factored into the formulation of policies and plans through the identification of development needs and initiatives, which are then assessed in the context of a vision for sustainable development.[7]

The SEA is conducted concurrently with the preparation of the draft management plan as illustrated in Figure 10.1. The SEA influences the zoning, plans, proposals, policies, and projects that are included in the final management plan, which will only be approved once the initiatives therein have been demonstrated to conform to both federal and Parks Canada policies, and to the intended use or protection of a particular zone. As projects are developed to meet the initiatives identified within the final and approved management plan, project level EAs will be conducted.

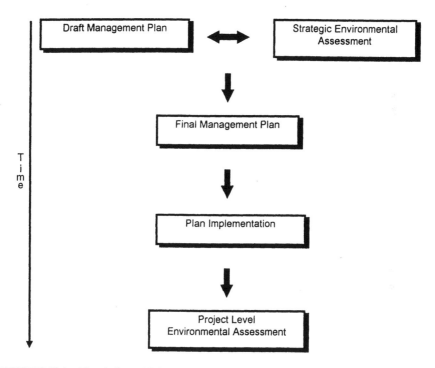

FIGURE 10.1   The timing of SEA in management planning.

PROCESS FOR CONDUCTING SEA

In conducting the SEA of a management plan, a multi-disciplinary team approach has been found to be the most effective and efficient means of conducting the assessment. The process begins with the assembly of a multi-disciplinary team, led by a team leader, who is responsible for coordinating the EA. The team leader will review the management plan with a view to identifying the various disciplines to be represented on the multi-disciplinary team. Typically, the team will consist of representatives from planning, natural resources, cultural resources, the socio-economic unit, visitor services, park or site management, and external stakeholders. The selection of representatives to participate in the SEA is a key factor in the quality of the assessment. All disciplines must be represented in order to ensure that the assessment is complete.

After the multi-disciplinary team has been identified, team members are individually tasked with reviewing the management plan to identify initiatives within the plan that may impact their particular discipline or interest. The initiatives arising from this individual review will include a combination of proposed or changed zoning designations, plans, proposals, policies, and specific projects. Although specific projects will mostly be conceptual in nature and will require further project-level EA or screening when detailed proposals have been developed, an SEA of a management plan can provide the framework for further EA requirements at a later date.

The next phase of the SEA is to conduct a screening workshop involving the representatives from all disciplines. The length of time required to conduct an EA workshop has been found to vary from one day for the management plan review of a historic site or park with few initiatives to two days for a more complex site or park. Several objectives are identified at the start of the workshop as follows:

1. to solicit agreement on the initiatives within the management plan which require environmental assessment
2. to identify a list of activities or events associated with each initiative
3. to assess the activities or events for impacts to heritage resources, both natural and cultural, using a variety of tools to assess project-environment interactions including matrices, tables, overlay maps, and geographical information systems
4. for each impact, to determine whether or not mitigation is appropriate, and if so, to identify possible mitigative measures, which could include modification or review of zoning, plans, or policies
5. to identify data gaps and determine the mechanism that would be appropriate for collecting or assembling missing information
6. to identify follow-up requirements, which may include monitoring to determine the effectiveness of mitigation
7. to assess the significance of the residual effects after the implementation of mitigation
8. to identify cumulative effects of the initiatives within the management plan

IDENTIFYING INITIATIVES

As mentioned above, the initiatives within a management plan that require assessment include plans, proposals, and policies which may set direction for a particular park or site for a number of years. Management plans for national parks and historic sites that have been assessed recently have included initiatives for scientific research, ecosystem, and cultural resource conservation, wildlife management, cultural resource management, access issues, operations, and recreational and economic opportunities.[10-15] The identification of initiatives is often complicated by the fact that initiatives may overlap. As an example, conducting scientific research may be identified as an initiative on its own merit, but research may also be identified as a component of resource conservation and management, and increased recreational and economic opportunity. Similarly, the development of research initiatives and increased recreational and economic opportunity may result in access and resource management issues. The selection of initiatives within the management plan that will need to be assessed will, therefore, be based on the major initiatives as recommended by the multi-disciplinary team, ensuring that all initiatives are captured and that repetition is avoided when initiatives overlap.

ASSESSING INITIATIVES FOR CONFORMANCE TO POLICY

The analysis of environmental effects resulting from an initiative begins with an examination as to whether or not the initiative conforms to Government of Canada environmental policies and Parks Canada environmental policies, guiding principles, and operational policies. The following questions should be addressed:

1. Does the initiative ensure and/or protect ecological integrity?
2. Is the initiative compatible with the commemorative integrity of the cultural landscape setting at a national historic site?
3. Will the integrity of cultural resources and their complexes be maintained?
4. Will basic scientific research and monitoring broaden the understanding of heritage resources and result in better management of heritage areas?
5. Are visitor activities designed to enhance understanding, enjoyment, and use of natural heritage while maintaining ecological and commemorative integrity, and protecting cultural resources?
6. Is the initiative consistent with the Government of Canada's directive respecting the greening of federal government operations as outlined in *A Guide to Green Government*?[16]

The goals and purposes of a management plan is that they must be environmentally sound and initiatives contained therein must be consistent with both federal and Parks Canada policies. For those initiatives that are not consistent with broader environmental policies, alternatives could be proposed or the initiative would be rejected during the SEA process.

ASSESSING INITIATIVES FOR CONFORMANCE TO ZONING

Once a determination has been made that the initiative is consistent with both federal and Parks Canada policies and principles, the next phase of the analysis is to determine whether or not the initiative conforms to the criteria established for the zone in which the initiative will be set. In national parks and marine conservation areas, the activities related to each initiative would be identified and assessed for conformance to the types of activities allowed for the zone in which the activity may occur. As an example, an initiative requiring or resulting in motorized access in zones designated as "Special Preservation" or "Wilderness" in a national park would not be appropriate, since access is restricted or prohibited. In a national historic site, park, or heritage canal, the initiatives would be evaluated for conformance with the designated zone, either contemporary or historical.

The process for producing a management plan within Parks Canada should generally preclude the advancement of initiatives to the final stages of the management planning process that do not conform with the zoning of a particular area. Boundaries and appropriate uses for the different zones of heritage places are determined by the planning team, at which significant interests and stakeholders are represented. Decisions respecting the direction and the initiatives in the management plan are based on scientific research, functional expertise, traditional and local knowledge, and extensive public consultation at the national, local, and regional levels. The evaluation of a proposal or initiative for its appropriateness with respect to zoning is, therefore, generally done within the context of the management planning process. The purpose of the SEA would be to confirm that the initiatives do, in fact, respect the zoning provisions in heritage areas.

Within the management planning process, there is a provision to allow activities which do not conform to the provisions for a zone in unique and exceptional circumstances. The justification to allow a non-conforming activity would be carried out in the context of management planning with the planning team and full public consultation. As a general guide, non-conforming uses would usually be short-term and guided by prescriptions that set out the allowable time-frame and any special circumstances that apply. An example of a non-conforming use in a national park is the issuance of grain farming leases in Grasslands National Park in areas which are actively being restored to native mixed grass as seed becomes available and resources permit. In the interim, farming is an effective short-term means of stabilizing the soils and discouraging the proliferation of exotic species. Since the complete restoration of these lands is expected to take approximately 15 years, grain leases are identified as non-conforming uses, which are fully justifiable in the interim.[17] The SEA of a non-conforming use would evaluate the need and the justification for the designation, recognizing that the activity, although contrary to zoning criteria, is short-term, may result in positive environmental effects or minimal adverse effects, will eventually be discontinued, and that no other alternatives are available.

If the SEA determined that an initiative was contrary to the zone in which it was planned, or that the need or justification for the designation of a non-conforming use was inadequate, the following courses of action would become available to the management planning team:

- The initiative could be abandoned.
- An alternative to the initiative could be considered
- A change in the park's zoning could be proposed.

Since management planning is conducted in a multi-disciplinary team approach with full public involvement, any changes to a draft management plan as a result of the SEA would necessitate public notice and full public participation in the consideration of a revised or new initiative, or a change in park zoning. The impact of the revised or new initiative, or of a zoning change in the draft management plan, would be re-assessed in the context of the SEA before the management plan was finalized and forwarded for ministerial approval.

If the SEA concludes that a proposal or initiative conforms to the zoning designation or is justified as a non-conforming use, an analysis of the possible environmental effects arising from the initiative would be undertaken.

IDENTIFICATION OF ANTICIPATED ACTIVITIES RESULTING FROM THE INITIATIVES

Following the assessment of the conformance with zoning for each of the initiatives proposed within the management plan, the next step is the identification of the activities or events resulting from the initiatives. As an example, several activities may be associated with an initiative that proposes to conduct scientific research in a national park including:

- control or removal of exotic species
- establishment of field camps
- access issues
- tagging, identification, and/or collection of flora and fauna
- monitoring activities
- excavation

The list of possible activities or events for each initiative is developed during the workshop by the multi-disciplinary team. Since the initiatives may be conceptual at this stage of planning and the details respecting the initiatives may be vague, the team must develop a reasonable list of activities or events that are based on the knowledge or past experience of the team members. The selection of the right team members, or follow-up with a specialist, is a key component in ensuring that the information generated during this exercise is complete.

ANALYSIS OF ENVIRONMENTAL EFFECTS

The next step of the assessment is to identify the potential environmental effects of each initiative. Environmental effects are defined within the Canadian Environmental Assessment Act as: (a) any change that the project may cause in the environment, including any effect of such change on health and socio-economic conditions, on physical and cultural heritage, on the current use of lands and resources for traditional purposes by aboriginal persons, or on any structure, site or thing that is of historical, archaeological, paleontological or architectural significance, and (b) any change to project that may be caused by the environment, whether any such change occurs within or outside Canada. Consistent with this definition of environmental effects, initiatives within the management plan are assessed for their impact on the biophysical and cultural environment, and on economic and social conditions.

The geographical scope of the assessment of effects arising from the initiatives within the management plan is not limited to the national park, national historic site, or marine conservation area, since it is recognized that these areas mandated by Parks Canada are not islands unto themselves. Parks Canada is committed to seeking mutually acceptable solutions when management practices result in transboundary concerns on adjacent privately or publicly owned lands. Parks Canada is also committed to cooperating with other levels of government, private organizations, and individuals for the planning of its adjacent areas to maintain ecological integrity within the larger ecosystem.

To identify the environmental effects arising from the initiatives, a variety of tools such as checklists, tables, matrices, modeling, map overlays, and geographic information systems (GIS) can be used. In conducting the analysis in a multi-disciplinary workshop setting, an effective mechanism has been to use matrices. The activities or events previously identified for each initiative are examined for possible effects to environmental attributes for each of the project phases, from construction and implementation to operation and final decommissioning. The matrix provides a quick "snap-shot" to focus discussion on the environmental effects. In the next phase, the environmental effects of each of the activities or events are identified and described in terms of the following:

- *Magnitude*: How severe will the impact be?
- *Frequency*: How often will this event occur — one time only or multiple occurrences?
- *Duration*: Will the impact occur over a short period of time or over an extended period?
- *Timing*: When will the event occur?
- *Geographic Extent*: What area will be impacted? Are transboundary effects possible?
- *Reversibility*: Are the environmental effects reversible?
- *Ecological Context*: Will the effect be in a valued ecological zone?

In conducting the SEAs of management plans, a written description of the predicted impacts used with the matrix has been found to be a useful method for organizing information. Again, the selection of the right team members, or follow-up with a specialist, is an important factor in ensuring that the prediction of effects is reasonable and complete.

MITIGATION OF ENVIRONMENTAL EFFECTS

Where possible, mitigation measures will be proposed by the multi-disciplinary team. Mitigation can be broad in scope and can include recommending modification or review of policy direction. Any such change would be subject to re-assessment before the management plan could be finalized as shown in Figure 10.1. In many instances, because of the conceptual nature of the initiatives and the lack of project-specific detail, proposed mitigative measures may be general in nature. As an example, the mitigation for an activity or event, which requires the construction of a trail, may be to comply with the *Best Practices for Parks Canada Trails*.[18] If possible, any surveillance during implementation of initiatives, or follow-up requirements to assess the accuracy of the analysis, will also be identified. Although it may not be possible to outline the specific details, general information as to the requirements and methods may be recommended.

The SEA can also recommend a framework for further EA as an initiative is developed from the conceptual stage to project development during the management plan implementation. An initiative which requires the development of a physical work by Parks Canada would trigger the Canadian Environmental Assessment Act. An EA at the project level would be conducted pursuant to the Act at a later date when details of the project become available and before irrevocable decisions have been made. Projects which are not subject to the Act may require an assessment pursuant to other relevant authorities as defined in the Parks Canada Management Directive 2.4.2.[3] It is appropriate to identify these further EA requirements in the context of the SEA.

ANALYSIS OF THE SIGNIFICANCE OF THE RESIDUAL EFFECTS

Residual effects are those effects which remain after the application of mitigation. Any initiative found to result in activities or events which have significant residual effects could not be recommended within a management plan. Significant effects are defined as effects that threaten ecological or commemorative integrity and are based on values such as loss of rare or endangered species, reductions in species diversity, loss of critical or productive habitat, transformation of landscapes, loss of cultural resources, and loss of current use of lands and resources for traditional purposes by aboriginal persons. If a conclusion was reached that an initiative would result in significant environmental effects, the initiative would have to be re-examined and modified by the planning team, and subject to re-assessment. The recommendation to approve a management plan is based on an analysis for environmental effects and a determination that the initiatives contained therein will not result in significant environmental effects, either individually or cumulatively. A management plan containing initiatives that could result in significant effects could not be recommended for approval.

There may be instances where there is inadequate information available to make a decision about the significance of a particular initiative. In this case, it would be appropriate within the SEA to identify the deficiencies and to recommend strategies to fill in the gaps. As an example, if an initiative promoted increased recreational use of a particular site, it may be reasonable to conclude that this initiative will not likely result in significant adverse effects. However, it may be prudent to recommend that surveys of use be conducted over a certain time frame, and based on this information, determine the carrying capacity of the affected area and establish limits if necessary.

CUMULATIVE EFFECTS EVALUATION

A final analysis of the management plan is for the combined or cumulative effects of the individual proposals. The cumulative environmental effects assessment evaluates the interactions between the initiatives within the plan and other projects and activities, both existing and foreseeable, and trends that are developing. Different heritage areas display different levels of impact, but all areas are subject to stresses, both internal and external, which can contribute to gradual loss of natural and cultural resources. Impacts can accumulate gradually through small insignificant increments over time and space. Since a management plan provides direction that attempts to balance heritage values and economic development, the cumulative effects assessment is a tool that can be used to establish the appropriate balance.

The process for conducting cumulative environmental effects has been described by Kingsley.[19] Kingsley recommends a four-step process: scoping, analysis, evaluation and follow-up, and feedback and documentation. When scoping, it may be appropriate to set different geographical or temporal boundaries for different environmental attributes. As an example, it may be appropriate to consider jurisdictional boundaries when evaluating socio-economic impacts, but more appropriate to consider a drainage basin for evaluating impacts to surface water. In conducting the evaluation, it is important to identify uncertainties and recommend, if possible, follow-up or feedback requirements to evaluate the accuracy of the cumulative effects assessment and the effectiveness of any mitigation proposed.

Maps and map overlays may be especially useful in identifying cumulative effects. As an example, in Prince Albert National Park, Saskatchewan, Canada, the cumulative ecological impact of development within the townsite of Waskesiu has been assessed by comparing the change over time using an energy classification system.[20] This information will enable management of the Park to set future development principles which reflect the requirement to not only maintain, but also to restore and enhance, the ecological health of the town site. A similar process of identifying and mapping surface and sub-surface impacts at St. Andrew's Rectory National Historic Site in Manitoba, Canada, will provide useful information on the past level of disturbance to cultural resources and their complexes for the upcoming management plan review. Any new initiatives which may result in further disturbances to surface and sub-surface cultural resources can be mapped to facilitate decision-making with respect to acceptable thresholds for further disturbances.

As stated by Kingsley, the cumulative effects assessment, therefore, becomes a key tool in a more holistic approach to planning based on temporal and geographical scales.[19] Assessing cumulative effects through the management planning process with its cyclical review can be a useful mechanism to provide feedback on past management decisions and guide future initiatives.

CONCLUSIONS

The SEA of management plans for heritage areas mandated by Parks Canada is a process by which the natural and cultural resources can be protected. Management decisions must be made applying cultural resource and ecosystem-based management principles. Initiatives proposed in the management plan can be assessed for environmental effects early in the planning process, before irrevocable decisions have been made, to ensure that there is a balance between heritage values and economic development.

A multi-disciplinary team approach in a workshop setting has been found to be an effective and efficient process to conduct the SEA. Simple tools such as matrices can assist in focusing discussion. The right balance of professionals from various disciplines including natural resources, cultural resources, site managers, socio-economic units, and external stakeholders will ensure that a broad range of experience and expertise identify the full range of activities and events, environmental effects, and mitigation required for an initiative that may be developed only conceptually.

The SEA also provides the opportunity to establish the framework for further EA needs, surveillance, and follow-up which may be required. Any critical deficiencies in information or uncertainties can also be identified, and strategies to address these can be recommended.

Finally, the SEA of a management plan enables the assessment of cumulative effects in a range of geographical and temporal scopes for different environmental attributes. Trends can be identified and assessed against different indicators including the ecological and commemorative integrity indicators. A better understanding of the state of the heritage areas within Parks Canada's mandate will be achieved over time.

REFERENCES

1. Canadian Heritage, *Parks Canada Guiding Principles and Operational Policies*, Minister of Supply and Services Canada, 1994.
2. Canadian Heritage, *Sustaining our Heritage – The Sustainable Development Strategy for the Department of Canadian Heritage*, Minister of Public Works and Government Services Canada, 1997.
3. Parks Canada, Management Directive 2.4.2, Impact Assessment, May 28, 1998.
4. Canadian Environmental Assessment Agency, *Strategic Environmental Assessment: The 1999 Cabinet Directive on the Environmental Assessment of Policy, Plans, and Program Proposals*, 1999.
5. Canadian Heritage, *Parks Canada Guide to Management Planning*, November 1994.
6. Sadler, B., *Environmental Assessment in a Changing World: Evaluating Practice to Improve Performance*, Minister of Supply and Services Canada, June 1996.
7. Partidário, M.R., Strategic environmental assessment: key issues emerging from recent practice, *Environ. Impact Assess. Rev.*, 16, 31, 1996.
8. Doyle, D. and Saddler, B., *Environmental Assessment in Canada Frameworks, Procedures and Attributes of Effectiveness*, Minister of Supply and Services Canada, March 1996.
9. Therivel, R., Wilson, E., Thompson, S., Heany, D., and Pritchard, D., *Strategic Environmental Assessment*, Earthscan, London, 1992.
10. Therrien-Richards, S., *Prince Albert National Park – Management Plan Environmental Assessment*, Technical Service Centre, Parks Canada, Winnipeg, December 1994.
11. Therrien-Richards, S., *Riding Mountain National Park – Management Plan Review Environmental Screening*, Professional and Technical Service Centre, Parks Canada, Winnipeg, June 1996.

12. Therrien-Richards, S., *Batoche National Historic Site – Management Plan Environmental Assessment*, Professional and Technical Service Centre, Parks Canada, Winnipeg, 1997.

13. Therrien-Richards, S., *Prince of Wales Fort National Historic Site – Management Plan Environmental Assessment*, Professional and Technical Service Centre, Parks Canada, Winnipeg, June 1997.

14. Therrien-Richards, S., *Aulavik National Park – Environmental Assessment of Management Plan*, Western Canada Service Centre, Parks Canada, Winnipeg, February 1998.

15. Therrien-Richards, S., *Wapusk National Park – Interim Management Guidelines Environmental Assessment*, Western Canada Service Centre, Parks Canada, Winnipeg, April 1998.

16. Government of Canada, *A Guide to Green Government*, Minister of Supply and Services, 1995.

17. Parks Canada, *Grasslands National Park Management Planning Program*, Draft Management Plan, 1999.

18. Canadian Heritage, *Best Practices for Parks Canada Trails,* Special Advisor's Office, Real Property Services (CH-EC) PWGSC, June 1996.

19. Kingsley, L., *A Guide to Environmental Assessments: Assessing Cumulative Effects*, Parks Canada, Department of Canadian Heritage, Ottawa, March 1997.

20. Thompson, R., personal communication, 1998.

11 Environmental Assessment and Planning in South Africa: The SEA Connection

Keith Wiseman

CONTENTS

INTRODUCTION

Strategic Environmental Assessment (SEA) has rapidly emerged as a tool for planning and environmental management in South Africa. Although some key questions regarding SEA processes and methods are yet to be answered, experience to date illustrates a number of benefits of a strategic approach to environmental issues.

This chapter describes how perceptions of Environmental Impact Assessment (EIA) have changed in recent years, leading to the emergence of SEA in South Africa. These changes have occurred in the context of changing legislation for planning which have shifted the focus of planning away from physical planning, toward the goal of facilitating appropriate and sustainable development.

Although no guidelines, protocols, or legislative requirements exist for SEA in South Africa, a number of SEA studies have been commissioned. Some selected SEA studies are presented in order to illustrate commonalities which exist and which characterize the various studies as "strategic." These studies also serve to illustrate the benefits which can result from SEA, as well as some of the problems encountered to date.

The commonalities and benefits, as well as problems associated with SEA, have important implications for an agreed approach to SEA in South Africa. These include links between SEA and planning, qualities which define the strategic nature of SEA and its relationship to project EIA, and the development of tools appropriate to SEA. The chapter concludes by emphazing the urgent need for an agreed approach to SEA in South Africa and identifies some implications of South African experience for SEA internationally.

A BRIEF NOTE ON TERMINOLOGY

The term "environment" is allocated to the same meaning in this chapter as it is in the South African guidelines for Integrated Environmental Management,[1] which, until 1998, established the framework for EIA in South Africa:

> The external circumstances, conditions and objects that affect the existence and development of an individual, organism or group. These circumstances include biophysical, social, economic, historical, cultural and political aspects.

THE CHANGING FOCUS OF ENVIRONMENTAL ASSESSMENT

Project-specific EIA has been in widespread use in South Africa since the 1970s. Although initially focused on the natural environment, social and economic impact assessment became common during the 1980s.

A milestone in the development of EIA was the publication of guidelines for Integrated Environmental Management (IEM) by the Department of Environment Affairs.[1] IEM provides a set of clear principles as well as a procedural approach to the consideration of environmental issues in the planning, decision-making, and implementation of development. The principles of IEM include public participation, accountability for information used, transparent decision-making, and ensuring that the social benefits of development outweigh the costs.

It was intended that the IEM procedure would complement planning activities, as well as provide a framework for project EIA.[2] In practice, however, IEM was seen by many planners as an unnecessary duplication of planning and decision-making procedures (for example, see Ref. 3).

Perceptions of duplication, together with the absence of legislative requirements for IEM, resulted in the application of IEM being often limited to project EIA In many ways, the terms "IEM" and "EIA" became synonymous. Few understood the differences between them, and the IEM procedure was rarely, if ever, implemented as part of planning.

The limitations of project EIA came to be widely recognized as a result of the Saldanha Steel Project (SSP). The SSP was a proposal to build a steel plant on the West Coast of South Africa, some 120 km north of Cape Town. A project EIA was undertaken and it identified a number of significant impacts and potential benefits.[4] However, the EIA was limited in its usefulness as a decision-making tool by a number of factors.[5] These included:

- The EIA was site-specific since the preferred site had been earmarked for industrial development since the 1970s and again in a 1993 Structure Plan.
- The assessment of secondary development and potential cumulative impacts was limited since developers preferred to delay an announcement of their plans until a decision to proceed with the SSP had been taken.

At least seven project EIAs were commissioned for industrial and related development in the Saldanha area following the SSP. This piecemeal approach to the environment is clearly limited in its capacity to assess cumulative impacts or address sustainability issues. In addition, the 1993 Structure Plan did not address potential environmental opportunities or constraints to industry or guidelines for project EIAs that would follow. This historical lack of environmental assessment (EA) in planning is discussed ahead.

THE NEED FOR SEA

The key issue in the above case was not whether the SSP could be constructed on the proposed site, but whether the Saldanha area was an appropriate location for heavy industry. Project EIA is poorly suited to this purpose. EA tools are needed, which can address issues at a level above the project-specific. In particular, EA tools are needed, which can assess and compare the capacity of natural and built resources, as well as social and economic networks, to sustain various development scenarios and alternatives.

Despite calls from objector groups that an SEA be undertaken before a decision on SSP,[6] no SEA was commissioned. At that time, many had heard of SEA, but few could claim to have seen it. Since 1995, however, SEA has emerged as a tool for the EA in the preparation of policies, plans, and programs (PPP).

THE CHANGING FOCUS OF PLANNING

Planning in South Africa has historically referred to "land use" or physical planning, and it has served to create and maintain development rights for property owners, reinforcing the policy of apartheid or separate development. However, South Africa's first democratic elections marked a turning point for planning.

BEFORE 1994

Past planning legislation focused on the spatial ordering of development as a result of two key concerns: to separate "conflicting" land uses and to protect private property rights.[7] Once approved, a Town Planning Scheme (TPS) entrenched development rights in perpetuity. No systems or procedures were in place to assess whether these rights were equitable or sustainable.

The first notable changes to this approach occurred in the mid 1980s. Planning legislation was introduced, which attempted to provide more flexible forms of planning that could respond to differing development needs as they arose. An example of this was the 1985 Land Use Planning Ordinance (LUPO) in the Cape Province. The LUPO established procedures for the preparation of Structure Plans, which neither gave nor removed rights, but which indicated proposed trends and types of development.

Procedures to involve the public in planning decisions were also prescribed by LUPO. Draft Structure Plans and any proposed rezoning of land were required to be advertised for comment. This system, however, contributed to the culture of "participation by objection"[8] in which public input was limited to objections late in the planning and decision-making process.

A further shortcoming of LUPO was the continued focus on land use. This resulted in a minimum of attention paid to broader environmental issues. For example, a Structure Plan could identify areas for development without consideration of the effect of development on stormwater quality, quantity, or downstream flooding.[9] Few, if any, EAs were used in the preparation of structure plans, and the plans themselves provided no guidance for project EIAs that might occur later.

AFTER 1994

Following South Africa's first democratic elections, the Government of National Unity undertook to rebuild planning legislation. A draft Development Facilitation Bill was soon published and public workshops were held country-wide to obtain comment on the draft. The Bill was finalized and became law in December 1995.

The Development Facilitation Act (DFA), therefore, represents the foundation for new approaches to planning throughout South Africa. It contains a number of principles to which plans must respond, providing for the first time substantive in addition to procedural guidelines for planning. The substantive principles established by the DFA include:

- Community empowerment in planning and development processes
- Addressing environmental concerns

- Speeding, planning, and development approvals in order to address back-logs
- Setting land development objectives (LDOs)

The restructuring of local government has occurred parallelly and was initiated by the passing of the Local Government Transition Act (LGTA). Newly elected local authorities are required by the LGTA to prepare Integrated Development Plans (IDPs). These are plans which integrate the financial, planning, and administrative functions of local government and include public participation in the identification of a vision and priorities for development. The spatial planning component of IDPs is provided by LDOs.

Both the DFA and LGTA therefore, advocate the fundamental principles of participation and sustainable development. Given these changes, the need for new approaches to EA was evident.

EMERGING FORMS OF SEA IN SOUTH AFRICA

Whereas both the DFA and LGTA identify the principles of community empowerment and concern for the environment, they do not prescribe methods by which environmental issues can be accounted for early in the planning process. However, during this transition phase, a rapidly increasing number of SEA studies were initiated. A selection of these is presented in Table 11.1 in order to identify commonalities that exist and the extent to which the studies are strategic. The selection of current SEA studies is intended to be representative of the emerging forms of SEA in South Africa.

A BRIEF DESCRIPTION OF THE SELECTED CASE STUDIES

Coega Harbour and Industrial Development Zone (IDZ)

This is a proposal to build a dedicated port and 10 km² industrial zone at Coega, some 20 km from Port Elizabeth in the Eastern Cape. The proposal is driven by a non-profit company representing local, regional, and national government, industry, and unions. The objective of the SEA was to determine whether or not the IDZ and port were feasible and, if so, under what conditions.

Durban South

Several large communities live alongside existing petro-chemical industries and the Durban Airport in the Durban South area. Public opposition to pollution, to project EIAs of the proposed expansion of industry and to plans to redevelop the airport lead to the initiation of the SEA. The goal of the SEA was to identify, assess, and compare alternative scenarios or visions for the future development of the Durban South area.

KwaZulu-Natal Regional Economic Forum

This regional economic forum was established in 1994 and tasked with developing trade and industrial development policy. The forum was representative of local and regional government, industry, and non-governmental organizations (NGOs). The SEA was commissioned by the forum to identify environmental opportunities and constraints, which should inform its proposed policies.

Cape Town 2004 Olympic Bid

Cape Town submitted its bid to host the 2004 Olympic Games to the International Olympic Committee (IOC) in September 1996. Since the city does not have all the infrastructure or facilities needed, a program for constructing and upgrading facilities was included in the bid. SEA was then used to assess whether the benefits would exceed the costs of bidding for and hosting the Games.

COMMONALITIES WHICH DEFINE THE NATURE OF SEA

The SEA projects presented in Table 11.1 show that both commonalities and differences exist between the various studies. These are significant to the development of an agreed approach to SEA in South Africa since no protocol or established framework for SEA yet exists. In particular, the successful development of SEA in South Africa requires that SEA studies be recognizably different from project EIA and avoid the perceptions of duplication as previously discussed.

The case studies reveal an emerging form of SEA in South Africa, which bridges the gap between project-specific EIA (Section 2) and the process of regional or sectoral development planning. In this way, SEA has the potential to guide regional or sectoral initiatives toward sustainability, whereas EIA is constrained by its delayed position and project-specific focus.

INITIATION OF THE SEA

In each of the case studies, it can be seen that key role players agreed on the need for the SEA. Without a screening mechanism to determine whether SEA is required,[10,11] consensus between stakeholders is perhaps the best foundation for the initiation of SEA. The key reasons for initiating SEA were:

- to provide input on environmental and sustainability issues to planning or decision-making
- to reduce the number and complexity of project EIAs
- to assess cumulative impacts and identify sustainability indicators

LINKS TO PLANNING AND DECISION-MAKING

The case studies show strong links between SEA and development planning. Three of the four studies were commissioned prior to, or as part of, a planning process. These SEAs were intended to facilitate appropriate and sustainable development by providing input to plans or policies. In the remaining case, the Olympic Bid SEA,

TABLE 11.1
Summary of the Key Features of Selected SEA Case Studies

	Coega IDZ and Harbour SEA	Durban South SEA	KwaZulu-Natal Regional Economic Forum SEA	Cape Town 2004 Olympic Bid SEA
Initiation of the SEA	The SEA forms part of preliminary feasibility studies to determine if the proposed industrial development zone and harbor were economically and environmentally sustainable.	Local communities rejected project EIAs due to problems associated with existing industries. Durban Corporation and affected communities eventually agreed on an SEA approach to be undertaken by a neutral third party.	The Forum recognized that the environment creates opportunities and constraints to increased trade and industrial development. An SEA was commissioned to provide input to the policy-making process.	The need to assess whether the environmental benefits would exceed the costs of hosting the 2004 Olympic Games was recognized by key role players including the Bid Company, NGOs, and local authorities.
Links to planning and decision-making	The SEA identified conditions, which must be imposed if the harbour and industrial zone goes ahead. These includ-ed guidelines for detailed planning, criteria for project EIA, guidelines for the environmental quality of air, water, and other resources, and systems for environmental management.	The SEA will provide input to future detailed planning, including guide-lines and limits to the expansion of industry. However, pressure exists to approve the expansion of some industries before the SEA is complete.	SOE analysis and GIS-based information systems have been completed for the policy Forum. While key role players expressed keen interest in the SEA, it is not clear how the SEA was integrated into policy, if at all.	Most venues and facilities were already chosen and priority projects went ahead while the SEA was undertaken. Never-theless, the express goal of the SEA was to inform a decision of whether to continue with the bid before the IOC announced its decision.
Approach and methods used in the SEA	Application of IEM principles, including public participation and review. Specialist studies of key bio-physical and socio-economic issues leading to proposed conditions under which the project could proceed. Capture and analysis of information using GIS.	Application of IEM principles, including public participation and review; SOE analysis of existing industrial processes and emissions. Identification of development scenarios and guidelines for future detailed planning.	GIS capture of data on existing industries, including inputs and outputs of raw materials and wastes. SOE analysis of air, water, and bio-physical opportunities and constraints. Socio-economic issues were not addressed	Application of IEM principles including public participation. Review of public policy to identify a "value system" to assess strategic costs and benefits. Baseline and predictive studies of key components of the affected environment.

TABLE 11.1 (CONTINUED)
Summary of the Key Features of Selected SEA Case Studies

Key products of the SEA	Guidelines for detailed planning, EIA, and environmental management. Sustainability indicators. Assessment of cumulative impacts.	Guidelines for detailed planning, EIA, and environmental management. Sustainability indicators. Assessment of alternative scenarios and potential cumulative impacts.	GIS-based information system. Framework for planning and assessing industrial development including updating GIS database.	Statement of strategic impacts and benefits according to a value system based on public policy.

GIS= Geographical Information Sytems.

the SEA was intended to be a decision-making tool, although it was too late to influence planning of the bid itself.

A further commonality amongst three of the case studies is the role of the SEA as part of a tiered approach to planning and environmental management. In the Coega, Durban South, and KwaZulu-Natal SEA studies, a clear goal of the SEA was the provision of guidelines, criteria, and sustainability indices for later detailed planning, project EIA, decision-making, and environmental management systems.

The use of SEA in a tiered system represents a new approach in South Africa and shows a willingness to address environmental issues at a considerably earlier stage than project level EIA. It is also consistent with emerging planning practice under the DFA and LGTA, in which priority issues are identified and addressed, leaving detailed or less urgent issues to be addressed later.

Approach and Methods Used in the SEA

A number of common methods emerge from the case studies:

- state of the environment (SOE) analyses based on specialist bio-physical and socio-economic inputs
- the use of GIS technologies to analyze data and create information systems
- public and authority participation in the identification and review of key strategic issues

These methods were used to a varying extent in each SEA to identify opportunities and constraints created by the natural and built (e.g., infrastructure) aspects of the environment and priority issues, which inform potential development opportunities. In the case of Durban South, these and other inputs from the public and stakeholders were used to develop alternative scenarios which could be tested against a range of goals and priorities. In this way, information resulting from the SEA studies provides a key input to the preparation of policies and plans as well as to decision-making.

KEY PRODUCTS OF THE SEAs

With the possible exception of the Olympic Bid SEA, the SEA studies have similar products and outputs. These include:

- GIS-based information systems
- guidelines for detailed planning, project EIA, and environmental management
- proposed sustainability indicators
- an assessment of possible cumulative impacts

In comparison, the Olympic Bid SEA produced a statement of the predicted strategic impacts and benefits that could result from hosting the Olympic Games. However, in 1997 the IOC awarded the 2004 Olympic Games to Athens.

CONCLUSIONS

It can be seen that SEA has emerged in South Africa as a result of three key forces:

- recognition of the limitations of project EIA
- the need for a form of EA which is integrated with planning processes and avoids perceptions of duplication
- new approaches to planning which address concerns for the environment, community empowerment, and a tiered system of development plans

These forces have influenced SEA in a way that appears to reinforce the changing role and focus of planning. SEA is needed, not only to address sustainability issues, but also because policy guidelines for project EIA and environmental quality have been absent from past planning practice.

THE STRATEGIC NATURE OF SEA IN SOUTH AFRICA

The needs described above require that SEA be recognizably different from project EIA. From the SEA studies presented, three characteristics of SEA can be identified, which define it as strategic as distinct from project-specific nature:

- SEA is applied at a higher level than the project-specific. Examples include regional planning, such as the Coega SEA, policy formulation, as in the KwaZulu-Natal SEA, and scenario development and testing, as in the Durban South SEA.
- The SEA forms part of a tiered approach to planning and the environment. Whereas a good project EIA should be comprehensive, SEA is focused

on priority issues, creating guidelines and criteria to address detailed issues in later stages.

- The focus of SEA is the effect of the environment on sustainable development, whereas a project EIA assesses the effect of a given development on the environment. The emphasis of SEA on *strategic environment* has a high potential to inform policy- and plan-making, promoting sustainable development, and ensuring informed decision-making.

Some Emerging Problems for SEA in South Africa

The rapid emergence of SEA in South Africa has not been without problems. In particular, five issues have emerged which must be addressed if SEA is to become a widely accepted tool:

- the lack of a screening mechanism to determine if SEA is needed
- SEA can take different forms since no agreed approach exists
- there are no legal requirements to ensure that the results or recommendations of an SEA are adopted and incorporated into plans or policies
- inadequate information on environmental issues, including a lack of baseline data and a lack of established tools to ensure that opportunities and constraints created by the environment are addressed in detailed planning and decision-making
- lack of strategic direction from key authorities

These problems are similar to those experienced in the early days of project EIA in South Africa. However, in the case of SEA, they are more complex and often intangible. For example, the lack of strategic direction from authorities illustrates both the need for SEA and one of the key problems associated with SEA approaches. In a worst case scenario, SEA could become a "lowest common denominator," with the word "strategic" added on to almost every form of EA. Similarly, the credibility of SEA could be severely undermined by just one or two poorly conducted studies.

SEA and Planning in South Africa: The Way Forward

Most SEA studies to date in South Africa have focused on the assessment of the environment and its effect on opportunities, constraints, and alternative scenarios for development. This approach is consistent with the role of SEA as a tool for planning and policy-making. However, for SEA to become widely established, both its relationship to planning and the substantive form of SEA must be clearly defined.

Until 1996, only the basic principles of SEA in South Africa had been described[12] and no systematic or substantive guidelines have been prepared. Further development of SEA has since included:[13]

- support for the development of SEA from key authorities
- participation of a broad range of role players from both the public and private sectors in the development of an agreed approach
- the identification of a set of substantive principles and tools which can guide the process while retaining the flexibility needed for the effective integration of SEA into policy- and plan-making

In addition, a White Paper on an Environmental Management Policy for South Africa[14] and a proposed new National Strategy for Integrated Environmental Management[15] have been published for comment. These documents address the need for new tools, including SEA, to form part of a tiered approach to sustainable development. The newly promulgated National Environmental Management Act (Act 107 of 1998) also makes provision for SEA to be legally required in various sectors and forms.

Notwithstanding these developments, the enactment of regulations for EIA in South Africa in 1997 has thrown into sharp focus the need for the integration of EA and planning. Planning approvals and EIA, for example, are administered by separate levels and departments of government, leading to even greater duplication and delays. SEA has the potential to remove these bottlenecks, but an agreed process and products are first needed if this is to be achieved.

SOME IMPLICATIONS FOR SEA INTERNATIONALLY

The need for SEA to be compatible with policy- and plan-making systems has been emphasized.[11,16] In South Africa, this flexibility has so far resulted in a focus on the strategic assessment of the environment, with less emphasis on the assessment of impacts. This focus, however, challenges current international definitions of SEA (for example, see Ref. 16).

Nevertheless, many of the problems and perceptions experienced in South Africa are similar to those reported for other countries. It is suggested here that the repetition of problems could be avoided in the following ways:

- There is a need to refine terminology. At present, SEA is a term which encompasses many forms of EA, leading to confusion and even contradiction.
- A range of EA tools and techniques is needed. An SEA, which assesses the environment as an input to policy-making, is different to the assessment of the strategic impacts of a proposed policy and its alternatives. The use of SOE analyses and the selection of environmental indicators are two techniques that have been used in the case studies presented in Table 11.1.
- Similarities and synergies between SEA and planning must be explored and analyzed in an interdisciplinary way.

The cry, *"But we already do that!"* is a common perception of planners in South Africa to SEA approaches. The tools, techniques, and established frameworks from the disciplines of both EA and planning must be shared if such perceptions — or misperceptions — are to be avoided.

ACKNOWLEDGMENTS

The author would like to acknowledge the support of the CSIR and the Netherlands Ministry of Foreign Affairs in the preparation of this chapter and attendance at IAIA '97.

REFERENCES

1. Department of Environment Affairs, *The Integrated Environmental Management Procedure.* Pretoria, South Africa, 1992.
2. Sowman, M., Fuggle, R., and Preston, G., A review of the evolution of environmental evaluation procedures in South Africa, *Environ. Impact Assess. Rev.,* 15:45–67, 1995.
3. Claassen, P., *Integrated Environmental Management as an Integral Part of Planning and Development,* Paper presented at the Institute of International ReS.E.A.rch Conference, Johannesburg, South Africa, January 31, 1996.
4. C.S.I.R. The Saldanha Steel Project Phase 2: Environmental Impact Report, *C.S.I.R. Report* EMAS-C94017C, Stellenbosch, South Africa, 1994.
5. Wiseman, K., *Balancing Theory and Reality in the Management of Large EIAs: The Saldanha Steel Project, South Africa,* Proceedings of the 16th Annual Meeting of the International Association for Impact Assessment, IAIA, Estoril Portugal, June 17-23, 1996, 681–686.
6. Steyn, J., *Report of the Commission of Enquiry into the Saldanha Steel Project,* Cape Town, 1995.
7. Milton, J.R.L., Planning and property, *Acta Juridica,* 267–288, 1985.
8. Sowman, M., *Socio-Economic Factors Affecting Estuarine Degradation: A Case Study of the Kaffirkuils Estuary,* University of Cape Town, Environmental Evaluation Unit Report, 4/88/27, 1988.
9. Wiseman, K., Stormwater management and land use planning in river systems, *Munic. Eng.,* 21(6), 22–31, South Africa, 1990.
10. Sadler, B. and Verheem, R., *SEA: Status, Challenges and Future Directions,* Ministry of Housing, Spatial Planning and the Environment, The Netherlands, 1996.
11. Partidário, M.R., Strategic environmental assessment: key issues emerging from recent practice, *Environ. Impact Assess. Rev.,* 1996:16:31–55, 1996.
12. C.S.I.R., Strategic environmental assessment: a primer, *C.S.I.R. Rep.,* ENV/S-RR 96001, Stellenbosch, South Africa, 1996.
13. C.S.I.R., *Towards Strategic Environmental Assessment Guidelines for South Africa: Division Document,* Stellenbosch, South Africa, 1998.
14. DEAT, *White Paper on Environmental Management Policy for South Africa,* Department of Environment Affairs and Tourism, Government Printer, Pretoria, South Africa, 1997.
15. DEAT, *Discussion Document: A National Strategy for Integrated Environmental Management in South Africa,* Department of Environment Affairs and Tourism, Government Printer, Pretoria, South Africa, 1998.
16. Therivel, R. and Partidário, M.R., *The Practice of Strategic Environmental Assessment,* Earthscan, London, 1996.

Section V

---

*Methods and Applications —
Role of SEA in Decision-Making*

12 Strategic EA and Water Resource Planning in Europe

Clare Brooke

CONTENTS

INTRODUCTION

Water resource issues are a high-priority concern in relation to wildlife conservation in southern Europe. A review of hydrological planning in the Tajo river basin, western Spain, was commissioned by BirdLife International to assess whether the process exhibited, or could exhibit, the principles of strategic environmental assessment

(SEA). Hydrological planning in Spain is based on a tiered-planning structure, and therefore, provided a good case study of whether a system of environmental assessment could be integrated at the different levels of decision-making.

WHY SEA?

BirdLife International is a global network of non-governmental organizations working for the conservation of birds and their habitats. The Royal Society for the Protection of Birds (RSPB) is the U.K. BirdLife partner.

Many BirdLife partners are influential in decisions made about land-use changes, which will affect birds or their habitats. They have found project Environmental Impact Assessment (EIA) to be indispensable in improving understanding of these land-use changes and highlighting their environmental implications. However, as in other cases, the EIA process has been shown to have its limitations, not the least of which is the fact that many project options are foreclosed as a result of decisions made prior to the project stage. It is this gap that BirdLife partners have sought we seek to close through the development of systems of SEA.

BirdLife International has been promoting SEA in several ways. A list of projects is given in the Appendix. Experience with these projects has confirmed the benefits of SEA as a flexible tool for assessing the environmental implications of proposals, and helping to promote environmental protection.

BENEFITS OF SEA

Listed below are some of the benefits that have been identified from work by BirdLife partners on SEA:

- SEA results in better decision-making. It allows environmental concerns to be taken into consideration earlier in decision-making, making it easier for damage to wildlife and habitats to be avoided or minimized.
- SEA helps to identify conflicting objectives within policies, plans, and programs (PPPs) and direct attention to prioritization of objectives and resolution of conflicts.
- SEA helps to identify and assign responsibilities for environmental protection to various authorities.
- SEA forms the context within which lower level assessments can be set. It should avoid the selection and advocacy of especially damaging projects, which, at worst, cause environmental damage and, at best, waste time and money.
- SEA can consider the impacts of measures, which are not implemented by specific projects.
- SEA allows alternatives to be properly considered.
- SEA enables the identification and analysis of cumulative impacts of several developments.
- SEA can provide baseline information for lower level assessments, and helps to identify data needs and gaps.

The above benefits have been rehearsed elsewhere (see for example, see Refs. 1 and 2). This chapter focuses on one of these benefits of SEA, that it provides a hierarchy or tiering of assessment, and forms the context within which lower level assessments can be set.

BENEFITS OF ASSESSMENT HIERARCHY

- A hierarchy of SEA and EIA focuses attention on the most appropriate stages to consider particular impacts.
- The procedures for SEA and EIA are similar (screening, scoping, alternatives, etc.).
- SEA streamlines the EIA process, which should consequently have a reduced scope and be easier.
- SEA provides baseline data for EIA and can reduce costs.
- SEA can clarify the environmental conditions for project approval.
- SEA can filter out the most damaging projects, removing the need for detailed, expensive, and controversial EIAs later on.
- SEA and EIA should give greater credibility and justification to decisions.
- SEA and EIA integration means that both strategic and project-level decisions should be based on a full assessment of their environmental implications.

One of the studies carried out by BirdLife examined this hierarchy in more detail.[3] The rest of this chapter focuses on this study.

SEA AND HYDROLOGICAL PLANNING IN THE TAJO BASIN, SPAIN

In 1994 BirdLife International commissioned a study to investigate how SEA principles could assist in resolving water resource planning problems. The study was based on hydrological planning in the Tajo river basin in western Spain. This examined the various levels of planning for water resources in Spain, and investigated whether elements of SEA existed within them.

CONTEXT

Water availability has been an important factor shaping the economic development of large parts of Spain. Water resource issues are also a high priority concern in relation to wildlife conservation. Water policy is, therefore, an issue of strategic and political importance.

Like other countries, Spain has acute problems with rising water demands, a series of severe droughts, and a vast disparity in rainfall between the north and south of the country. Spain is the largest water consumer in the European Union (EU) and the fourth largest in the Organization of Economic Co-operation and Development (OECD).[4] The main reason for this lies in the poor efficiency of water transportation and irrigation, and the inadequate policy of water pricing.[5] Agriculture accounts for

65% of total demand and this continues to rise. This demand sector is primarily concentrated in the southern half of the country.[5] To increase water availability in large parts of the country, more than 1000 large reservoirs have been built.

Hydrological planning in Spain has the potential to cause significant environmental damage. For example, major water infrastructure developments are planned, which will have severe impacts on the natural environment. Nationally, 600,000 ha of new irrigation and over 200 new dams were are planned. A total of 164 areas identified as internationally Important Bird Areas (IBAs) would be are affected.[6] IBAs are areas that qualify for designation as Special Protection Areas (SPAs) under the EU Wild Birds Directive 79/409/EEC. Twelve of these are already designated as Ramsar sites under the Convention on Wetlands of International Importance and 50 are designated as SPAs.

The problem of water resource provision is one of the most complex and controversial problems in Spain. BirdLife International wanted to investigate whether SEA principles could be applied to the hydrological planning system in Spain to help avoid or reduce environmental conflicts. The study provided a challenging subject for an analysis of SEA.

Hydrological planning in the Tajo basin in western Spain was chosen as a case example. The Tajo basin extends from east to west, draining an area of 55,645 km², extending into Portugal. It has over 200 dams and around 703 km of irrigation channels. It is also a very rich area for nature conservation. Much of the area is farmed extensively with sheep and cattle, enabling rich grasslands to develop. It also has much of the remaining *dehesa* habitat of cork and holm oak and olive groves.

Hydrological planning in Spain is defined by an extensive statutory framework. This forms a hierarchical structure (see Figure 12.1). At present, the planning of the use and enhancement of water resources is carried out through national and regional instruments. The general context is set by the Spanish Constitution, which gives the state responsibility for the overall planning of economic activity. National legislation in the Water Law defines the objectives, content, and scope of hydrological planning. Below this, the detailed instruments for hydrological planning are the National Hydrological Plan and the individual catchment plans. Each of these has been given a specific role, that of the National Hydrological Plan (PHN) being to coordinate the various individual catchment plans. The national plan also has the power to determine policy for water transfers between catchments. Catchment plans subsequently provide the context for specific projects.

Each of these levels are briefly considered, in turn, with an analysis of whether they exhibit elements of SEA.

THE WATER LAW

The Water Law passed in 1986 by the Central Government contains a number of key conservation objectives, which appear capable of underpinning a sound system of water resource management. The Law defines the objectives of planning as:

> to satisfy water demand to the best possible extent and to ensure balanced and smooth regional and sectoral development by increasing water availability, protecting water

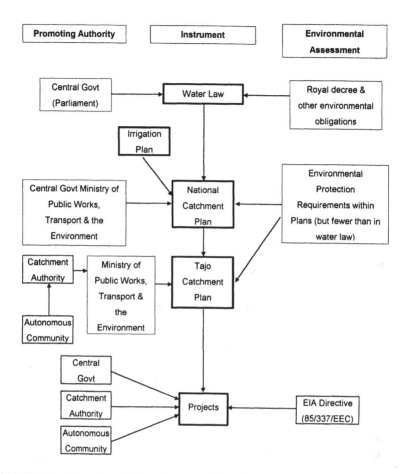

FIGURE 12.1 Hierarchy of Water Resources Planning in Spain. (From BirdLife International, *Strategic Environmental Assessment and Hydrological Planning in Spain,* BirdLife International, Cambridge, U.K., 1995. With permission.)

quality, ensuring sparing water consumption and reasonable water use in sympathy with the environment and other natural resources.

However, there is no apparent priority ranking afforded to these objectives, many of which are in clear conflict.

The Water Law also contains particular environmental objectives, such as establishing protection zones, providing resources for conservation, and implying an environmental minimum flow to protect biodiversity.

The Water Law and Environmental Assessment

The Water Law provides the broad context for SEA. It includes environmental objectives, which could be assessed, and requires plans, programs, and projects undertaken by the government to be assessed.

Royal decree 849/1986 (Regulation on the Public Water Domain) accompanying the law also provides explicitly for EIA of all applications for use of water, where the catchment authority believes such use could risk pollution or environmental degradation.

THE NATIONAL HYDROLOGICAL PLAN

The National Hydrological Plan is prepared by central government and approved by Parliament. Its general objective is "to meet current and future demand by means of the judicious use of each catchment's own resources and balancing water availability between catchments." It is the first attempt at defining such objectives for water resource planning in Spain.

It also includes a further series of objectives, including:

- to increase water supplies
- to guarantee water quality for each use and for environmental protection
- to protect people and land from flooding
- to conserve hydrological infrastructure and historic monuments

The national plan identifies the main problems that need addressing as the uneven geographical distribution of water supply and the lack of appropriate charging. It proposes means of addressing these through inter-basin water transfers, better monitoring of water flow, and charging for water volumes actually used.

It then outlines the methods for achieving its basic objective of "satisfying demand." These are *indirect* through water conservation and re-use; and *direct* through increasing supply by providing infrastructure. The national plan gives priority to the latter, arguing that "indirect methods, although less costly, would not of themselves satisfy demand." It proposes a complex system of engineering works connecting all catchments and allowing water to be transferred to wherever it is thought to be appropriate (the Integrated National Water Redistribution System - SIEHNA).

There is a clear contrast between the early objectives, which include emphasis on resource protection and efficiency, and the ultimate conclusions, which concentrate on increasing supply almost as an objective in itself.

The National Hydrological Plan and Environmental Assessment

The National Hydrological Plan includes some elements of environmental assessment (EA). It sets out the procedure for preparing catchment plans. These must identify resources, current and future use, infrastructure needs, quality, and resource protection requirements. These components allow an overall resource management strategy to be designed and a long-term balance between demand and supply attempted. So the plan contains elements of analysis which are akin to SEA.

However, the existing system includes flaws, which prevent adequate SEA being carried out. For example, there is no attempt to identify key priorities among conflicting objectives. Also, the actual measures proposed in the national plan are dominated by the basic objective of satisfying demand. For example, a complex system of engineering works is proposed, connecting all catchments and allowing water to be transferred to wherever its needed. This creates problems lower down the system in individual catchment plans because the environmental impacts of these transfers are not considered.

There is also no requirement in the national plan for EA other than for those projects listed under the project EIA Directive 85/337/EEC. There is no reference to SPAs for birds (under Directive 79/409) even in those cases where these have been formally designated. The environmental impact of proposals on these internationally important habitats has therefore not been evaluated.

THE TAJO CATCHMENT PLAN

Catchment plans are prepared by Catchment Authorities (who are autonomous agencies of central government). They require approval by the Government after a request from the Water Council of the catchment concerned. Catchment level planning follows a consistent framework laid out in the PHN.

The Tajo basin was selected for detailed study for the following reasons:

- The basin contains a good representation of ecosystem types and IBAs.
- Hydrological planning in the catchment poses direct threats to wildlife and habitats. Fourteen IBAs are likely to be affected by the proposed irrigation schemes, and four IBAs by the proposed dams.
- The Tajo basin plan was one of the best studied of the hydrological plans.
- The Tajo authority was described as the most advanced, a "flagship" for hydrological planning in Spain.
- Several types of planned developments are represented in the basin, each with substantial water demands.
- Seventy-four percent of water consumption is by agriculture, mainly for extensive irrigation. Another major use is electricity generation. There are plans to increase resources further, primarily for supply to Madrid. Measures include water transfers, further exploitation of surface and groundwater, and the construction of fifteen new dams.

The Tajo plan contains similar provisions as the National Hydrological Plan. The environmental elements of the plan appear to have been weakened further, instead of the expected "devolved" environmental responsibilities re-emerging. For example, the only site protection guidelines are for wetlands within the catchment. There is no mention of other sites such as SPAs, or sites protected under regional legislation. None of the wetlands cover those of artificial origin, such as reservoirs, although 11 such sites in the catchment are identified as IBAs. The areas selected for protection appear to be precisely those with which planned works will not interfere.

The Tajo Plan and Environmental Assessment

The Tajo plan does include the requirement for an "environmental" minimum flow. However, this is treated as an operational condition rather than a water use constraint, and so it is not considered as important as demand uses. The constraints in the plan focus on the rivers, and not on the wetlands, which may be dependent on them.

The Tajo plan has clear recognition of the need for economic appraisal of "project" options. However, the study found that the environmental implications of the Tajo catchment plan had not been rigorously considered. As a result, 17 IBAs were threatened. Other wildlife will obviously suffer as well. The areas of threat include important habitats such as *dehesas*. Wooded dehesas are pasturelands populated by holm and cork oaks. They have an understorey of open grassland, cereal crops, or Mediterranean scrub, most commonly with a typical savannah appearance. This kind of ancient agricultural/pastoral system is exclusive to the western Mediterranean basin, and supports a high number of endangered bird species both in spring and winter, such as the Spanish imperial eagle and the great bustard. It also supports a wide diversity of other plant and animal species, particularly rich flower grasslands.

The nature conservation interest of these areas has been maintained by the traditional extensive mixed farming in the dehesa. However this is now increasingly under threat from intensification of farming, encouraged by subsidies for mechanization, but also by improved irrigation, allowing more crops to be grown. The environmental consequences of this intensification, and the role of irrigation, need to be considered through a proper assessment of the catchment plan.

The other threat to habitats comes from the construction of new dams, 15 of which are proposed for the Tajo basin. Further schemes are proposed, which will threaten habitats, potentially causing adverse impacts on birds, such as wintering Cranes. Lynx are also affected by dams, which flood valley floors where they like to feed.

ANALYSIS OF SEA AND HYDROLOGICAL PLANNING IN SPAIN

The hydrological planning system in Spain, therefore, contains some of the elements of EA, but the BirdLife study concluded that it fails to deliver a comprehensive analysis of the environmental impacts of water resource developments. Most of the assessments built into hydrological planning were related to EIA rather than SEA, focusing on detailed assessment of projects rather than a more strategic assessment of policies and plans. Hydrological planning continues to be driven by the need to meet demand, especially for irrigation.

Many of the environmental requirements of the Water Law have not been fully transferred into the more recent National Hydrological Plan or Tajo Catchment Plan. Only the water quality monitoring and waste water control requirements are carried through fully. Other environmental requirements are either lost because of different interpretations (for example, guaranteed flow to conserve aquatic biodiversity) or because of a transfer of statutory responsibility (for example, designation of protection zones).

A major problem identified was that the Water Law, National Hydrological Plan, and even the Tajo Catchment Plan, all include conflicting objectives. The lack of prioritization between these objectives, and failure to resolve the conflicts, has meant that environmental objectives have not been met, and potential environmental impacts have not been properly assessed.

Each level of decision-making includes measures for environmental protection, but these are not carried through into the planning process, often because of the priority given to increasing supply. Consequently, the lower tiers of plan have not provided a forum for considering long term, strategic decisions, but have instead, simply provided the framework for management of water resources, with the objective of achieving long term water supply.

Part of the difficulty in achieving the environmental objectives related to the failure to properly assign responsibilities for environmental protection to authorities. Poor data on existing environmental resources was a serious constraint, and consequently, the analysis of impacts was inadequate. Full data availability is not a prerequisite of SEA, but the process of conducting a SEA will identify gaps in data or knowledge which need to be filled at lower levels before final decisions are made. This process can also lead to the identification of monitoring requirements for the plan, and any data collection needed to inform the reviews of the plan.

Lack of data also contributed to the failure to consider alternatives, and there was no attempt to consider alternative options and choose the least damaging. Finally, there was poor adherence to national or EU environmental legislation such as the Birds (79/409/EEC) and Habitats and Species (92/43/EEC) Directives.

However, the BirdLife study concluded that SEA *could* be incorporated into hydrological planning within existing policy frameworks, and that it could achieve positive environmental benefits. SEA would provide baseline information, scope out the most damaging projects, and reduce the need for costly and detailed EIAs at the project stage. The tiering of decisions from the Water Law, cascading down into the different plans, provides an opportunity for similar tiering of assessment, starting with a strategic assessment of the Water Law and National Hydrological Plan, and a more detailed assessment of the options in the catchment plan, feeding down into the EIA of specific projects implementing the catchment plan.

In order to streamline the assessment process, it is important to ensure that the appropriate level of detail is used at each stage, becoming progressively more detailed as one moves down the level of decision-making. The sorts of assessment needed at each stage are given in Table 12.1. This is not comprehensive, but it highlights how the hierarchy of assessment might work.

The BirdLife study attempted to identify how an analysis of predicted impacts could be streamlined through the hierarchy. At the national level, the assessment should focus on the number of European or national designated nature conservation sites affected by the national plans, particularly proposals for major infrastructure projects such as dams and irrigation. Then, at the catchment level the assessment would consider the extent of the potential impacts and their magnitude, considering all wildlife sites. It should also look at cumulative impacts. Further down the line at the project level the effect on all wildlife resources, including direct and associated impacts would be considered. This, therefore, provides a consistent structure of assessment at each level of decision-making, and helps to ensure that the environmental implications are explicit, and ideally, damage is avoided.

TABLE 12.1
The Scope of a SEA System

Issue	National Hydrological Plan	Assessment Activity Tajo Catchment Plan	Projects
1. Environmental objectives	Fulfill EU and national obligations. Maintain environmental minimum flow in all rivers/wetlands.	Fulfill EU, national and regional obligations. Maintain environmental minimum flow. Restore wetlands.	Fulfill Eu national, regional, and local obligations. Maintain/enhance key species.
2. Character of the area	Extent of SPAs/Special Areas of Conservation in each catchment. Extent of nationally important sites.	Boundaries of EU/national/regional sites. Red Data Book species inventory.	Boundaries of natural habitat types. Inventory of key species present.
3. Strategic options	Water conservation. Inter-basin transfer. Agricultural reform.	Leakage control. Irrigation control. Major scheme options.	Alternative location, magnitude, design.
4. Predicted impact	Number of European/national sites affected by dams, irrigation.	Impact extent and magnitude on all sites. Cumulative impacts.	Effect on wildlife resource including indirect and associated impacts.
5. Constraints identified	Requirements of Water Law. Requirements of Habitats Directive.	Requirements of National Hydrological Plan/EU funding/Habitats Directive.	Requirements of National Hydrological Plan/EU funding/Habitats Directive.
6. Selection of option and rationale	Emphasis on leakage. Agricultural reform to reduce need for new schemes.	Reduction in overall new water supply. Reduced abstraction to assist conservation of wetlands.	Amendment to project to safeguard key habitat/species.
7. Mitigation proposals	Major schemes amended or dropped. Inter-basin transfers reduced.	Major schemes amended or dropped. Wetland creation.	Amended design. Habitat enhancement.
8. Implications for next tier	Catchment plans must reduce amount of irrigation. Drop particular scheme?	Projects must address indirect impacts.	May be implication for operational activities.
9. Monitoring and review	Review of National Hydrological Plan. Database of national/international nature conservation sites established.	Review of Tajo Catchment Hydrological Plan. Water level monitoring in key wetlands.	Field monitoring of construction and operational phases.
10. Uncertainties	Not all Special Areas of Conservation identified.	Not all IBA boundaries defined.	Seasonal difficulties of survey.

Source: From BirdLife International, *Strategic Environmental Assessment and Hydrological Planning in Spain, BirdLife International*, Cambridge, U.K., 1995.

The result of SEA, as an iterative process, should be to influence the choice of options according to their predicted environmental impacts. This may include amendments to the planned supply of water; to the balance between measures to increase water efficiency and new resource development; and to the extent of permissible inter-basin transfer. In the Tajo plan, it will point to implications for the general locations of new infrastructure. A major failure of the Tajo plan was the lack of any assessment of cross-boundary impacts, since the downstream part of the Tajo catchment is in Portugal. Decisions made in the catchment plan, particularly to increase water supply for irrigation and the construction of dams, will have significant implications for the rest of the catchment. These implications should be considered in the plan, particularly given the requirements of the Espoo Convention on transboundary impacts.

SEA would identify the substantial environmental impacts of the current plans, and would help to highlight the gaps in environmental protection which exist in the system at present. SEA could ensure that revisions to each of the items of legislation and planning built in environmental safeguards in a comprehensive and effective way, without undue overlap. It would assess a range of alternative strategies and options for water resources management, and provide consistent information upon which to base choices. End projects could then be developed within a more environmentally sustainable framework. An essential component of SEA is public involvement (or at least consultation), and the plan preparation process would need to be modified to incorporate opportunities for such involvement.

This BirdLife study demonstrates the usefulness of a hierarchical approach to EA. If the Tajo plan had been strategically assessed, in a truly iterative way, environmental constraints could have been built in. Detailed project assessment will continue to be essential. However, it may be able to proceed more efficiently and with less controversy, since environmental principles would have already been established. This, of course, depends on whether the results of the SEA were taken into account in decision-making and influenced the selection of more environmentally acceptable options.

There is one danger with this hierarchical idea as strategic decisions may still need to be made on the basis of fairly broad brush assessments. It is possible that projects will be selected on this basis, which subsequent detailed EIA shows to have serious unforeseen effects. Project EIA is, therefore, still important to ensure that damage to wildlife and habitats does not occur. In the context of the Tajo study, which concentrates on strategic issues, the continuing importance of the final project EIA should perhaps also be emphasized, especially in the absence, as yet, of these other tiers of appraisal. Project assessment, with good scoping, consideration of alternatives, well-designed mitigation, and post-project monitoring, must work effectively if the end results are to be environmentally acceptable.

WAYS FORWARD

THE SPANISH CONTEXT

Events since the BirdLife study was completed offer the hope that elements of a more strategic approach may yet develop. The Spanish Parliament did not approve the National Hydrological Plan in 1994 as anticipated because of the lack of information on irrigation. A National Irrigation Plan was adopted in 1996, establishing a maximum of 190,000 ha of new land to be irrigated (compared to the 600,000 ha originally planned). A new government was elected in 1996 and the National Irrigation Plan was again brought into question. Further studies are being carried out, and a revised Plan is likely.

The Spanish BirdLife partner (Sociedad Española de Ornitología) subsequently commissioned further research into SEA,[7] to investigate whether a methodology could be developed for SEA of EU Structural Fund programs, since these have an important influence over whether proposals for expanded irrigation in Spain go ahead. The study investigated the system in Castilla y Leon, the only region in Spain to have SEA legislation. A 54% increase in irrigated land is proposed for the region. Again, this will have significant implications for the region's wildlife.

The study focused on the Duero catchment, and found that implementation of the irrigation and catchment plans would result in a decline in the number and distribution of 63% of the bird species studied, including the Spanish imperial eagle (endangered at the international, European, and national levels). It would also have negative impacts on steppe habitats. The study concluded that the plans would result in the region failing to meet its other objectives, such as the regional government's model for sustainable development, requirements under the EU Habitats and Birds Directives, and also agricultural production restrictions. It provides an interesting example of how SEA cannot only highlight the environmental consequences of plans and programs, but also identify conflicting objectives between them.

The Castilla y Leon and Tajo studies illustrate that SEA principles can be integrated into existing planning systems, and would have many benefits. However, they also show that there are many obstacles in achieving such a system. This situation is repeated all across Europe. One of the ways to solve this problem is to establish a European Directive on SEA.

EUROPEAN UNION DIRECTIVE ON SEA

Context

The European Commission always intended to introduce a comprehensive system of EA, covering Policies, Plans, and Programs (PPPs) and projects. However it was recognized early on that this could not all be achieved at once. Consequently, the EIA Directive (85/337/EEC) was introduced initially. It has taken much longer than hoped for mechanisms to be agreed upon for the other assessments. In 1993, the Commission submitted a report to the European Parliament on the implementation of the EIA Directive. This concluded that many EAs were restricted because most of the decisions affecting projects were taken earlier in the planning process. Consequently, there was little opportunity to consider alternatives or mitigation measures at the project stage. As a result, new impetus was given for the assessment of plans and programs. In December 1996, a proposal was adopted by the European community for a Community Directive on the assessment of the effects of certain plans and programs on the environment (the SEA Directive). This has been subsequently amended and is awaiting approval.

Hierarchy of Assessment

The SEA Directive is intended to complement the EIA Directive and provide a hierarchy of assessment from plans and programs to projects. The rationale for the SEA Directive is that it will produce a more efficient assessment system. It will mean that the appropriate type of assessment will be required at the relevant stage in the decision-making process. Under the proposed Directive, strategic issues will be assessed at the plan and program level, leaving the EIA at the project level to address specific issues inherent to the proposed project. By conducting a comprehensive assessment at the strategic level, parts of the information required by the EIA Directive can be provided, resulting in a more streamlined assessment at the project level.

Early drafts of the SEA Directive provided an excellent rationale for the need to treat EIA and SEA in a hierarchical manner. This was based on four principles:

- Each assessment should be closely related to and be integrated with, the action and stage within the planning process to which it applies.
- The content should cover those aspects best dealt with at the given stage.
- The assessment should sit logically with those at previous higher and subsequent lower stages. There should be neither omissions nor unnecessary duplication.
- The level of detail and precision should normally reflect the stage in the planning process to which it relates.

The proposal SEA Directive also recognizes the importance of the hierarchy, although it does not set out the principles as clearly. It sets out the information required in an environmental statement, which should take into account "the level of detail in the plan or program, its stage in the decision making process and the extent to which certain matters are more appropriately assessed at different levels in that process."[8]

Potential Influence of a SEA Directive in Spain

The Tajo study provided an example of how it could be possible to integrate SEA and EIA into water resource planning. The difficulty in this case was that it failed to transfer environmental objectives in national legislation through to implementation. A EU Directive would help to improve hydrological planning in Spain in several ways (the text in parenthesis indicates the relevant provisions of the proposed SEA Directive):

- SEA would be required for the National Hydrological Plan and individual catchment plans, such as the Tajo, but not the Water Law (Article 2).
- Some of the environmental protection elements could be strengthened as the Directive will require European designated sites and other environmental protection requirements to be considered as part of the assessment (Annex).
- Environmental responsibilities would have to be assigned to an environmental authority (Articles 5 and 6).
- Clear objectives would have to be set and alternatives considered (Annex).

The European Directive would, therefore, improve environmental decision-making in relation to water resource use in Spain. However, as currently drafted, the Directive could be improved in certain critical ways:

- It should apply to policies as well as to plans and programs.
- It should apply to European Community actions.
- It should apply to marine activities.
- It should require monitoring.
- It should require deficiencies in data to be identified.

This would enable an integrated system of EA throughout decision-making hierarchies, from high-level strategic policy assessments through to detailed project assessments.

CONCLUSIONS

Governments worldwide are now committed to sustainable development. The aim should be to ensure that principles of SEA are integrated into the overall decision-making processes, and guide those decisions.

EA should follow a logical hierarchical and temporal sequence of appraisal of PPPs and projects. This will ensure consistency of approach and comprehensive coverage of issues, while avoiding unnecessary duplication.

Studies (for example,[9,10]) have shown that the benefits of SEA depend on the level at which the strategic assessment is taken:

- The greatest benefit is gained if there is a tiered environmental policy context, which sets out the environmental aims. SEA can then be used to test whether PPPs are consistent with these aims.
- Benefit is gained to a lesser extent if SEA of a given PPP is carried out to ensure that possible damaging elements are avoided or mitigated, and positive gains are maximized, and that the projects that arise from the PPP reflect this.
- As a minimum, SEA can assess or review the environmental impacts of a PPP so that damaging impacts can be mitigated and beneficial ones enhanced.

Finally, one of the greatest benefits of SEA identified in some studies[9] is that it informs those who are carrying out the SEA. It is, therefore, vital that those preparing the PPP also carry out the SEA (perhaps with an independent verification to ensure quality of assessment). This is one of the major differences with EIA, where the assessment is often carried out by external consultants or experts. The long term vision and benefit of SEA should therefore be to change the culture of organizations, governments, and individuals, so that environmental considerations are at the heart of all decision-making. As such, SEA is one of the key tools for working toward sustainable development.

REFERENCES

1. Thérivel, R., Wilson, E., Thompson, S., Heaney, D., and Pritchard, D., *Strategic Environmental Assessment,* Earthscan Publications Ltd, London, 1992.
2. Thérivel, R. and Partidário, M.R., *The Practice of Strategic Environmental Assessment*, Earthscan Publications Ltd., London, 1996.
3. RSPB/SEO BirdLife Espana, *Strategic Environmental Assessment and Hydrological Planning in the Tajo Basin, Spain,* Summary Report of a study by Lola Manteiga and Rodrigo Jiliberto, RSPB, Sandy, 1995.
4. *Almanaque Mundial*, Pub. América Ibérica, 1993.
5. Dirección General de Obras Hidráulicas, *Memoria del Plan Hidrológico Nacional* Ministerio de Obras Públicas y Transportes, 1993.
6. Grimmett, R.F.A. and Jones, T.A., *Important Bird Areas in Europe*, International Council for Bird Preservation, Cambridge, U.K., 1989.
7. SEO BirdLife Espana, *Developing a Methodology for the SEA of Regional Development Programs: The Case Study of Hydrological and Irrigation Plans in Castilla y Leon, Spain,* SEO BirdLife Espana, Spain, 1998.
8. European Commission, Amended Proposal for a Council Directive on the Assessment of the Effects of Certain Plans and Programs on the Environment, com (96)511 and com(99)73, 1999.
9. Bedfordshire County Council/RSPB, *A Step by Step Guide to Strategic Environmental Appraisal,* Bedfordshire County Council, Bedford, U.K., 1996.
10. Bina, O., Briggs, B., and Bunting, G., *Towards and Assessment of Trans-European Transport Networks' Impact on Nature Conservation*, BirdLife International, Cambridge, U.K. 1997.

APPENDIX

BirdLife International Work on SEA

Thérivel, R., Wilson, E., Thompson, S., Heaney, D., and Pritchard, D., *Strategic Environmental Assessment,* Earthscan Publications Ltd, London, 1992.
RSPB/SEO BirdLife Espana, *Strategic Environmental Assessment and Hydrological Planning in the Tajo Basin, Spain,* Summary Report of a study by Lola Manteiga and Rodrigo Jiliberto, RSPB, Sandy, 1995.
Bedfordshire County Council, RSPB, *A Step-by-Step Guide to Environmental Appraisal,* Bedfordshire County Council, Bedford, 1996.

BirdLife International, *Methods for Assessing the Ecological Impact of the TENs: Technical Workshop 24-25 April,* BirdLife International, Cambridge, 1997.

BirdLife International, *Strategic Environmental Assessment and Corridor Analysis of Trans-European Transport Networks,* BirdLife International, Cambridge, 1996.

BirdLife International, *The Community Support Framework for Greece 1994-1999,* BirdLife International, Cambridge, 1994.

BirdLife International, *The Community Support Framework for the Objective 1 Regions of Spain 1994-1999,* BirdLife International, Cambridge, 1994.

BirdLife International, *The Environment and the Structural Funds: The Role of Strategic Environmental Assessment – the Sicilian Experience,* BirdLife International, Cambridge, 1997.

13 The Use of SEA and EIA in Decision-Making on Drinking Water Management and Production in the Netherlands

Rob Verheem

CONTENTS

1-56670-360-3/00/$0.00+$.50
© 2000 by CRC Press LLC

INTRODUCTION

Environmental Assessment (EA) for decision-making on drinking water management and production, at both strategic and project level, has been mandatory in the
Netherlands since 1987*. EAs have been carried out for strategic decision-making
at the national level (in particular, for decisions on *sources* and *methods* of drinking
water production), for the *siting* of drinking water production in specific regions,
and to determine the *amounts* to be abstracted at specific sites. This chapter discusses
the tiering of Strategic Environmental Assessment (SEA) and Environmental Impact
Assessment (EIA) at different levels of drinking and industrial water management
planning, the methodologies used, and the effects on decision-making in three case
studies at different administrative levels:

- the SEA for the National Plan on Drinking and Industrial Water
- the EIA for the selection of a site for a deep infiltration project in a
  sensitive coastal dune area to the west of the city of Amsterdam
- the EIA to determine the most environmentally friendly way to produce
  drinking water from two wells on an ecologically valuable island off the
  north coast of the Netherlands

TIERING: EA FROM POLICY TO PROJECT LEVEL IN THE WATER SECTOR

Relevant decisions on water management are taken at both strategic and project
levels. To carry out EA at only one level would not be sufficient. Therefore, SEA
and EIA is carried out at several levels in the Netherlands. This, however, creates
the risk of overlap and double work, if the same issues would be dealt with at several
levels. To prevent this, it is important to carefully identify which sort of assessment
should take place at which level and how these relate to each other. In the water
sector this has led to the following situation.

In the Netherlands, as in most countries, in the planning process from the national
to the regional and local level at some point the following four questions have to be
answered. Why do anything? What must be done? Where must it be done? How
must it be done? The *why*-question deals with the need, objectives and principles
of new actions. Once the need has been established, the *what*-question deals with
selecting the best methods and the capacities needed for each of these methods. The
where-question is about the location of facilities, installations, etc. The *how*-question

---

* See Ref. 1 for an overview of the SEA and EIA process in the Netherlands.

deals with topics such as the detailed design of projects, necessary mitigation measures, and compensation issues.

In the water sector in The Netherlands the *why*-question and part of the *what*-question (preferred methods for water production) are dealt with in the National Plan on Drinking and Industrial Water. The remaining *what*-question (capacities needed) and the *where*-question are formally dealt with in the private water sector's Ten Year Water Plans. For these plans no SEA is carried out. The *how*-question is dealt with in the licensing process for specific sites, and for these, an EIA is carried out if the project is above certain thresholds*. In Box 13.1 an overview is given of the current planning process and how SEA and EIA are integrated in this process.

Although formally dealt with in the Ten Year Plans many regional water bodies choose voluntarily to include *what*- and *where*-questions in their EIAs for license applications. In practice, therefore, most major environmental questions are addressed by an EA in the Netherlands. Nevertheless, in the formal planning process a gap appears to be in the application of EA, i.e., insufficient mandatory coverage of the *what*- and *where*-questions. Currently, the planning process is being revised. It is not yet clear what the new planning system will look like and how EA will fit into it, but one of the ideas is to concentrate most major decisions in the Company Plan of the regional water bodies. For this plan an integrated (S)EA is then carried out. Principles and long term objectives will then be set in national plans such as the National Environmental Policy Plan, possibly requiring the application of the new environmental test (E-test).

Box 13.1: EA in the Dutch drinking water management planning process

National level

Why do anything and what must be done (methods)?

Need		
Objectives	National Plan on Drinking	SEA carried out by
Principles	and Industrial Water	responsible ministries
Methods		

Regional level

What must be done (capacities) and where must it be done?

Locations	Ten Year Water Plan	No SEA required
Capacities		

Local level

How must it be done?

Design		
Mitigation	licensing process	EIA by proponent
Compensation		(regional water body)

---

* Capacity more than 5 million cubic meters per year in sensitive areas, 10 million cubic meters in other areas, water reservoirs with a surface more than 100 ha and main water pipes with a length more than 10 km.

SEA FOR THE NATIONAL PLAN ON DRINKING AND INDUSTRIAL WATER

ISSUES

The two main goals of this SEA were to determine the ecological impacts of alternative national water production policies and to compare environmental and other aspects of alternative methods of water production.[2]

ALTERNATIVE POLICY OPTIONS

As a first step in the assessment, five alternatives for future national water production policy were developed. Two broad categories may be distinguished:

A - Using Existing Production Methods:

- increasing total drinking water production
- reducing total drinking water production
- reducing the industrial use of water

B - Altering Production Methods:

- increasing the existing use of ground water (shallow and deep ground water, infiltrated river water), decreasing abstraction from surface water
- reducing current use of ground water, increasing the use of surface water

Assessment Approach

The environmental effects of the alternatives were assessed in three steps:

1. The development of national hydrological models (for both ground water and surface water) and an appropriate geographic information system. Using these models and prognoses of the future water production capacities needed in each of the alternative policy options, the impacts of alternatives on surface water and ground water in the Netherlands were then determined.
2. The development of a model to determine existing natural values of moist and wet ecosystems in the Netherlands (the DEMNAT model). The main features of this model are the identification of homogenous ecosystems ("ecotope groups") and the estimation of the existing natural value of these ecosystems per square kilometer, based on:
   - the presence of ecotope groups
   - the national and international rarity of these groups
3. Determination of changes in existing natural values expected as a result of the influence of the various policy alternatives on the state of surface water and ground water.

Results

The approach described above produced the following results:

- There is a *direct relation* between the level of drinking water production and ecological impacts.
- Ending all *ground water* abstraction would lead to a 12% increase in the natural value of moist and wet ecosystems (compared with 1988).
- Ending all *drinking water* production would lead to a 10% increase in natural value.
- Ending all *industrial use* of water would lead to a 2% increase in natural value.
- Ending abstractions from shallow ground water would be most effective in raising natural values, followed by deep ground water, infiltrated river water, and industrial use.

ALTERNATIVE PRODUCTION METHODS

The SEA made a comparison of production methods:

- use of ground water: shallow ground water, deeper ground water, and infiltrated river water
- use of surface water: direct abstraction via a natural reservoir and an artificial reservoir
- use of artificial infiltration: surface infiltration and deep infiltration

Assessment Approach

The following approach was taken:

1. The following environmental aspects were compared:
   - nature effects
   - landscape effects
   - effects on the abiotic environment: use of resources, waste production, energy

In addition to environmental aspects, public health, use of space, and technical/economical aspects (such as availability, flexibility, vulnerability, and costs of methods) were assessed.

2. Several *subcriteria* were defined for each aspect.
3. A *mix* of quantitative and qualitative information provided the basis for scoring each of the subcriteria.
4. Scores for subcriteria were translated into *one score* using a mix of methods (normalization).

5. Sensitivity analyses were carried out.
6. For each aspect, methods were classified from "best" to "worst" on the basis of a multicriteria analysis (MCA), with weights reflecting different perspectives: health, abiotic environment, nature, landscape, and economy.

Results

The main conclusions from each of the perspectives were broadly the same:

- best score: deep ground water, infiltrated river water, and deep infiltration
- medium score: surface infiltration and natural reservoir surface water
- worst score: direct extraction from surface water, shallow ground water, and artificial reservoir surface water

SEA Quality Review

The Commission reviewed the SEA and considered the quality to be good. In particular, the development of the DEMNAT model was judged favorably. However, the lead authority was advised to adopt caution when applying the results of the assessment at the regional level. The production techniques that score best in the SEA could perform differently in the regions due to specific hydrological situations in each case (water abstraction does not affect nature in all regions) and/or developments in related sectors within a region, such as agriculture. For example, it would not be very effective to end the abstraction of ground water for drinking water production in a specific region if this meant that the same water would later be used and discharged to surface water by farmers (for example, to improve soil structure to allow the use of farm machinery). The Commission advised the selection of a framework of EIA measures aimed at the conservation or development of nature (related to water production).

Effect on Decision-Making

According to the competent authority, the SEA did influence the decision-making process. The results of the SEA were taken into account when formulating national policy for future public water infrastructure in the Netherlands. Furthermore, the methods developed as part of the SEA both stimulated and structured project EIAs in the water sector, which facilitated interpretation of the National Plan when preparing plans at the regional level.

EIA FOR THE LOCATION OF DRINKING WATER PRODUCTION CAPACITY

ISSUES

Government policy in the Netherlands is to move away from the use of shallow ground water and surface infiltration to the use of deep infiltration of surface water.

This was also one of the outcomes of the SEA described above. The present use of shallow ground water is especially harmful in sensitive nature areas, such as the coastal dune area where it leads to desiccation. An EIA was carried out to identify the best site for deep infiltration in the Overveen Coastal Dune Area (west of the city of Amsterdam).[3]

ALTERNATIVE SITES

Alternative sites to be examined further in the EIA were found by screening all potential areas on the basis of geomorphological and planning constraints, in particular existing landforms and land uses such as houses, campsites, etc. Forty potential sites remained.

Assessment Approach

The best areas for drinking water production were identified using an MCA. In this MCA, the 40 potential areas were scored on 7 aspects:

- ecology
- visual impacts/landscape
- technology/hydrology
- recreation
- agriculture
- effects on living area
- financial costs

Parameters were defined for each of these aspects. As an example, the following parameters used for ecology and landscape are listed below:

Ecology:
- the site's potential for the development of ecologically valuable ecosystems, both aquatic and terrestrial (moist dune valleys) ecosystems
- the possibility of using existing infrastructure
- the need for leveling, both on and outside the site (e.g., for pipelines)
- existing valuable vegetation on the site
- existing valuable fauna on the site
- existing degree of disturbance on the site (e.g., recreation, eutrophication)

Landscape:
- possibilities for camouflaging any necessary buildings on the site (e.g., relief, vegetation)
- visual effects of removing trees and shrubs
- possibilities for integrating necessary buildings into the landscape

As no quantitative information was available for most of the parameters, it was decided to score all parameters on the basis of expert judgement. A three point scale was used: a site is scored as being "relatively positive" (score 1.0), "neutral" (score 0.5) or "relatively negative" (score 0) for each parameter. On the basis of these scores, the sites were ranked from "best" to "worse" with the use of MCA. In calculating final rankings, weights were given to scores to reflect different priorities. Three sets of weights were used to reflect technical priorities (the quantity and quality of the water), environmental priorities (environmental protection), and a mix of concerns (a compromise between technical and environmental priorities).

Results

From the EIA it could be concluded that from all perspectives:

- the sites outside the dune area score best
- within the dune area, one specific site scores best
- the site originally preferred by the water company scores poorly

The sensitivity analyses that were carried out showed the above conclusions to be sound.

EIA Quality Review

In its review, the Dutch EIA Commission concluded that the assessment was of good quality. The EIA provided all information necessary for further decision-making. However, the Commission also concluded that it was not clear how the information in the EIA had played a role in the preparation of the license application submitted to the competent authority for which the EIA had been carried out. This application proposed choosing the site originally intended by the water company, despite the fact that this site scored poorly in the EIA.

Effect on Decision-Making

In its final decision, the competent authority decided not to select any sites outside the coastal dune area. The main reasons for this were the absence of available water infrastructure and the inevitable problems associated with property rights. The site in the dune area originally preferred by the water company was eventually chosen in combination with the site that scored best on environmental aspects in the EIA.

EIA FOR THE AMOUNT OF DRINKING WATER PRODUCTION

ISSUES

Currently 150,000 m³ of water is abstracted each year from the deep ground water under the dunes in the middle of the island of Schiermonnikoog, which lies off the north coast of The Netherlands. In the future (2015), the demand for drinking water is expected to increase to 230,000 m³ per year. The existing rate of abstraction already causes damage to valuable moist and wet dune ecosystems, which is why the possibility of extracting water from a well outside the dune area was investigated. This well is situated close to the southern shore where there are no dunes. The EIA was carried out to determine whether ground water abstraction at the new site is more environmentally friendly than at the old site. A second question was whether it would be better to abstract all the water needed in the future from the well outside the dune area (which entails certain risks to the quality of the water) or use a mix of water from the old and the new well.[4]

ALTERNATIVE PRODUCTION QUANTITIES

The following alternatives were assessed in the EIA:

- A - maximum abstraction of 125,000 m³ per year outside the dune area and a maximum of 75,000 m³ inside the dune area, with a constant proportion being provided from each well
- B - the same amounts, but the amount abstracted inside the dune area is kept constant while the amount abstracted outside the dune area is varied to match peaks in demand during the summer
- C - the same amounts as A, but flexibility is sought in the amount abstracted inside the dune area while the amount outside the dune area is kept constant
- D - all the water needed is abstracted outside the dune area; the well inside the dune area is closed.

Assessment Approach

The alternatives were assessed on the following parameters:

- ecological value of dune valleys
- surface water quantity
- chemical quality of surface water
- impacts on birds
- impacts on aquatic vegetation
- landscape/cultural history
- public health/safety
- risk of accidents
- purification
- construction/maintenance

The main elements in the assessment were the impacts of alternatives on ground water levels, surface water levels, the quality of surface water and ground water (hydrochemical effects), and impacts on vegetation and (avi)fauna. The following relation was assumed:

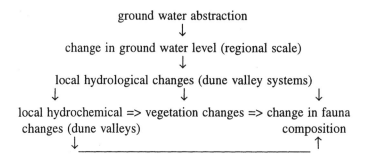

A ground water model was used in the EIA to make quantitative determinations of the changes in ground water levels at a regional scale. These calculations, in combination with known hydrological characteristics of the area, were then used to make qualitative assessments of local impacts (hydrological and hydrochemical). Subsequently, impacts on vegetation and fauna were determined qualitatively using an approach developed at the University of Groningen. Field observations were used to relate hydrological and hydrochemical changes to the presence of indicator species and groups of species. A qualitative assessment was chosen because insufficient ecological and hydrological knowledge was available for a quantitative assessment. Impacts on birds were also determined qualitatively, based on expected vegetation changes and existing knowledge of the preference of birds for certain types of vegetation structure. All other criteria mentioned were discussed qualitatively on the basis of expert judgement. The impacts were finally presented in the EIA on a seven-point scale.

Results

The EIA showed that Alternative D (all water abstraction outside the dune area) yielded the most beneficial environmental outcome. In this alternative, ground water is abstracted directly before flowing into the Wadden Sea, and thus, plays no further role in the ecology of the dune system. However, in this alternative, the surrounding area should be irrigated to prevent it from drying out, especially in the summer. The EIA also showed that of the three alternatives involving a combination of water abstraction inside and outside the dune area, Alternative C scored the best on environmental aspects.

EIA Quality Review

In its review, the Commission found the Environmental Impact Statement (EIS) to be of good quality. Although the assessment of ecological effects could have been more specific, the EIS contains all the necessary information to allow the competent authority to fully take the environmental issues into account when making a decision.

Effect on Decision-Making

On the basis of the EIA, the competent authority decided to grant a license for water abstraction as described in Alternative C, on the condition that within five to ten years, all water abstraction will take place outside the dunes (Alternative D).

CONCLUSION

The case studies show that a methodology and tools are available to carry out effective and influential EAs of drinking water facilities at all levels of decision-making. In the cases described this led either to a final decision in line with the best option for the environment or to a decision in which the environmental impacts were balanced against the technical and financial issues.

REFERENCES

1. Tonk, J. and Verheem, R., *Integrating the environment in strategic decision making — one concept, multiple forms*, Paper to IAIA98, Christchurch, 1998.
2. Ministry of Housing, Spatial Planning and the Environment, *SEA National Plan on Drinking and Industrial Water (MER Beleidsplan Drink en Industriewatervoorziening)*, Den Haag, 1993.
3. Grontmij, *EIA Deep Infiltration Overveen (MER Diepinfiltratie Overveen)*, De Bilt, 1990.
4. Iwaco, *EIA Drinking Water Production on the Isle of Schiermonnikoog (MER Drinkwaterwinning Schiermonnikoog)*, Groningen, 1994.

14 SEA of the Neiafu Master Plan, Vava'u, Tonga

Richard K. Morgan and Komeri R. Onorio

CONTENTS

INTRODUCTION

The Kingdom of Tonga comprises three main island groups across approximately 600 km of the South Pacific (Figure 14.1). The main northern group, the Vava'u group, is home to just over 15,000 Tongans, most of them living on the main island, 'Uta Vava'u, whose largest settlement is the town of Neiafu (population approximately 5,000). Neiafu sits overlooking a large natural harbor, Port of Refuge, well known in yachting circles for its beauty and safe anchorage in the hurricane season. The outer islands of the Vava'u group are mainly coral atolls, and the American

1-56670-360-3/00/$0.00+$.50
© 2000 by CRC Press LLC

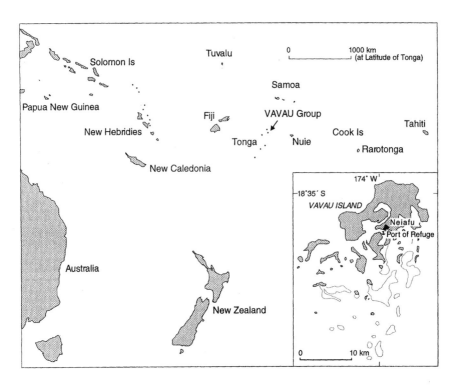

Figure 14.1   Location of Vava'u, Tonga.

author, Paul Theroux, in his book, *The Happy Isles of Oceania: Paddling the Pacific*[1]
typifies the reaction of many western visitors to this island group:

> We were sitting on the dock at the Port of Refuge, among the perfect little islands of
> Vava'u, each of which was a perfectly rounded piece of land....simple little places with
> no people---that was the thrill, the innocence of it, and anyone with a little boat could
> play Robinson Crusoe here. Each one was just what you imagined a tropical island to
> be---palms, woods, surf on the bright beach, limpid green lagoons. I was so glad I had
> come, and felt that I had discovered an island that few others knew, and had found a
> way of going there and living on it. That was the realization of the South Seas dream...

For the people of Vava'u, of course, the reality is different. Tonga is a developing
country, heavily reliant on multilateral and bilateral funding of development projects,
particularly infrastructure and economic development initiatives, to help improve
the living conditions for its people. The economy of Vava'u is very much based on
fishing and agriculture, with remittances and tourism also making important, though
lesser, contributions. Not surprisingly, development projects in Vava'u have been
targeted at improving the infrastructure and services that support agriculture, fishing,
and tourism in the region.

   In this chapter we describe a strategic environmental assessment (SEA) carried
out in 1996 of a development plan[2] proposed for the town of Neiafu, as part of the

wider Vava'u Development Program, funded by the European Union (EU) under the Lomé Convention. The plan contained a number of projects for improving local living conditions, and local shipping facilities for the agriculture and fishing sectors. However, the plan also envisaged major enhancements of the townscape to improve Neiafu's tourism potential, and this reflects a basic tension in the plan between the needs of the local people and the development of the tourism potential of the island group, a theme we discuss later in the chapter.

THE PURPOSE OF SEA

Usually SEA is promoted as a tool for use during policy, program, and plan (PPP) development, with the findings of the SEA being fed back into the development process to ensure biophysical and social impacts are being recognized, and wider aims, such as sustainable use of resources, are being met. However, SEA is not yet widely practiced in such a way, and it is often the case that the development process has progressed as far as a set of proposed projects before explicit environmental assessment (EA) is called for (perhaps not even then, in the worse cases).

In Tonga, we were faced with such a situation, but because the Tongan Government sought the advice of the Environmental Impact Assessment (EIA) Officer at the South Pacific Regional Environment Program (SPREP) (the officer being, at that time, Komeri Onorio, co-author of this chapter) on the assessment of certain projects proposed in the Neiafu Master Plan, it provided an opportunity to suggest that an SEA be carried out on the Plan as a whole. The combination of all the proposed activities in the Plan would constitute a major change in the character of the town and its role, not only in Vava'u, but in the country as a whole. Vava'u supports about 20% of the Tongan population and Neiafu is its principal center; any substantial change in the economic, social, and cultural character of Neiafu is of major regional and national importance. Equally, unintended consequences of the proposals, particularly if they affect the economic and social base of the local area or the region, will be of great importance.

A particular benefit of SEA is that it allows a coherent overview of a set of proposed activities that might possibly have effects on the same geographical area and/or the same environmental sectors, and the Neiafu Master Plan, comprising a number of discreet projects, required this type of scrutiny. SEA is also particularly useful when a proposal has been evolving over a number of years, as it provides an opportunity to step back and review the evolving PPP and to ensure the basis of the proposal is environmentally sound. Again, the Neiafu plan fell into this category, its origins going back to the early 1980s.

The SEA process, in common with all EIA activities, is a fundamental tool for the promotion of sustainable development, as it emphasizes the need to think through the environmental consequences of proposed developments before committing to a proposal. In that way, environmental costs and trade-offs can be recognized explicitly in the decision-making process and steps taken to enhance long term sustainability of the final development. Accordingly, a major theme in our assessment of the Neiafu Master Plan was the degree to which concepts of sustainable development were represented in the policy- and plan-making processes, and in the Master Plan itself.

One of the key tenets of SEA is the need to safeguard the natural resource base of local communities, hence it is very important that decision-makers, as well as potential resource developers, have a clear understanding of the carrying capacity of the local environment.

For the purpose of this investigation, the environment was defined as including not only biophysical components, but also the human communities affected by the proposals; consequently, the lifestyles, values, health, and culture of the local people (broadly summarized as their quality of life) are seen as part of the overall environment that might be changed. In terms of the concept of sustainability, the support of local communities is given a high weighting, which makes the avoidance of adverse impacts particularly important. Another major issue in planning for sustainable development is the importance, where possible, of not reducing options for future resource users. This means that careful thought must be given to the consequences of proposed actions to ensure their impacts do not unintentionally constrain such future options.

There are no provisions for SEA under Tongan legislation at present. Indeed, the government of Tonga is still working on legislation to introduce project-level impact assessment into the resource development process. To the best of our knowledge, this study is still the only SEA to have been conducted in Tonga, and given the institutional situation, this might not change in the foreseeable future. However, one purpose of this project was to demonstrate the benefits of SEA, regardless of whether it is required by statute. Our hope is that SEA will be used on a voluntary basis if need be, in recognition of the benefits to policy and plan development.

METHODOLOGY

The SEA was approached in three ways:

a) The background to the Neiafu Master Plan, its links to wider policy initiatives for Vava'u and for Tonga as a whole, and the general environmental context of the plan, were examined. General features of the PPP development process, could then be assessed for their degree of adherence to good environmental planning criteria, particularly the notions of sustainable development of resources.

b) Next, the component activities making up the plan were examined in a broad, but coherent fashion, using a simple impact (interaction) matrix. This gave some feel for: the totality of the environmental impacts (i.e., a cumulative assessment); the relative distribution of impacts between the biophysical and social environments; and specific issues that had not already been identified in relation to individual projects in the Master Plan. In some cases, more detailed environmental impact assessments were obviously required before projects could be allowed to go ahead.

c) In the light of the broad, potential, environmental implications of the development plan, we examined the issues of monitoring and the feedback of information into the development activity. The emphasis here was on monitoring for the wider environmental concerns, rather than project specific concerns, although there is a necessary link in reality.

Our investigation was based on the 1993 plan as it stands, but subsequently the Tongan Government has responded to the plan by giving priority to certain activities and downplaying others. At the end of the chapter we briefly comment on the recent thinking on the development issues in the Vava'u Development Committee (VDC) and Vava'u Development Unit (VDU). It should also be noted that the time available for the study imposed certain constraints on the SEA. In particular, we were not able to talk with people involved in the earlier stages of the process, before the Master Plan was published. Instead we had to base our assessment and evaluation on the official documents produced by the process. As such, our conclusions have to be somewhat guarded.

Another constraint on methodology is the attitude to public participation in Tongan public affairs. Normally, it would be expected that an SEA of a town plan would involve local people, to ensure their opinions, concerns, and values, were represented in the assessment process. However, there is not the same recognition of this type of public involvement in Tonga, and it was not possible, nor indeed advisable, to try to arrange for local community input for the study, especially as the "field work" took place over a comparatively short time period (ten days were spent in Vava'u).

POLICY CONTEXT OF THE NEIAFU MASTER PLAN

Vava'u Regional Development

The Neiafu Master Plan is the product of a planning process which began in the early 1980s, but which was given greater focus in 1986 with the establishment of the VDC. The purpose of the committee is to plan for the social and economic development of Vava'u, and since its establishment it has commissioned a number of reports to this end. An important document was the so-called "Atkins" report, or the *Vava'u Regional Development Program, Phase II: Final Report, Project and Program Dossiers*,[3] completed by the consulting firm WS Atkins International and published in August 1989. In Phase I, the consultants had examined the social and economic character of Vava'u and made a number of recommendations for development projects aimed at the improvement of the economic base of the region and improving living conditions of local communities.

The Phase II report outlines specific projects, making up an overall development program for Vava'u. The Executive Summary of the Atkins report (p. 1-2) describes the program as:

A strategy to promote the primary sector and maximise natural advantages.

The development strategy is based on a balanced program which reflects the rural lifestyle of the population and the predominance of the primary sector in employment and wealth creation. It also recognizes that the scenic and environmental quality is such that Vava'u has significant tourist appeal, a potential which must be harnessed through a sensitive and phased development programme, which conserves and enhances these assets.

And later in the Atkins report (Section 2.2.a), by way of background information:

The programme is designed to enhance rural employment and income levels through active promotion of export crops and improved marketing of fish products. Enhanced tourist attractions are intended to increase visitor expenditure and hence incomes in the region whilst at the same time conserving the natural heritage and environmental balance of Vava'u. Road improvements will be influential in promoting agriculture and tourism facilities and hence are an important part of programme design. The central town of Vava'u, Neiafu, requires a number of improvements in order to raise the standards of living for its inhabitants and these considerations have been assessed fully. The overriding requirement is clearly for a Master Plan of the town to be undertaken in order to assist future planning. Social considerations have been made (sic) in assessing health and education requirements.

The Neiafu Master Plan then is a direct product of this process. The Atkins report provided a reasonably detailed brief of the topics to be addressed in the plan:

1) building conservation and regulations
2) town amenity sites
3) transport management
4) potable water supply
5) urban sewage disposal
6) power station
7) new market site
8) harbor/wharf area
9) strategic planning of foreshore and the old Neiafu harbor

Overall, the Master Plan was to guide future development, limit land use conflicts, and improve architectural and design standards. The Plan itself was produced in 1993 and presented as an integrated plan for the development of the town, incorporating the specific activities listed above. A technical feasibility study was released in 1994 and it proposed more detailed investigations of a number of the projects described in the original plan.[4]

The VDU, established in 1992, is charged with implementing the projects contained in the Plan, together with development projects in other areas of Vava'u, such as jetty development and solar-powered lighting in the outer islands. Projects linked to the Master plan included drainage improvements in the town center, wharf redevelopment, and re-location of the open market. Other projects will be undertaken as the program develops and funding is finalized. The Tongan Water Board has a national program underway for improving water supplies, and the Neiafu water supply problems are being examined within that program.

NATIONAL DEVELOPMENT PLANS

The whole Vava'u development initiative has to be seen in the wider context of national development policies. The Tongan National Plan, produced on behalf of the Government by the Central Planning Unit, represents a periodic appraisal of the social and economic state of the country and sets out national development objectives for the various sectors of the economy and for social services such as health and education. The planning for the development of Neiafu has taken place under the fifth and sixth National Plans and, as a consequence, reflects the broad economic and social objectives of the central government.

The Sixth National Plan[5] does not explicitly consider the environmental implications of the development objectives set out in the Plan. However, in 1990, the United Nation's Economic and Social Commission for Asia and the Pacific (ESCAP) produced an Environmental Management Plan for Tonga, which does address many of those issues.[6] The ESCAP document reviews the major sectors of the Tongan economy and identifies important environmental impacts and issues associated with those sectors. The document also contains a summary of the institutional arrangements for managing the environment and dealing with the problems raised by the various economic activities, from farming and forestry to fishing and tourism. This is used to identify possible improvements to the environmental management framework for the country.

COMMENT

The Atkins report set the context for the Master Plan in clear terms. The focus was to be on improving the living standards of local people by using the available natural resources, but clearly it also recognized the need for wise use of those resources, be they agricultural, fishery, or aesthetic resources. The implication is clear: the resulting development proposals were expected to balance the economic, social, and environmental needs of the region so that local people could achieve a better standard of living, while protecting their resource base. This is one of the main principles under-pinning the concept of sustainable development and it is clear that the Neiafu Master Plan was expected to sit within this context.

Three observations can be made, however. First, it is not clear from the Atkins report to what extent the local communities in Vava'u participated in the development program to that stage, and there is no indication that participation ought to be a consideration in the continuing program. Certainly, the VDC has two People's Representatives to the Legislative Assembly (that is, Ministers of Parliament (MPs)), but there appears to have been no other formal mechanism for seeking the views and concerns of the community. It is widely recognized that major initiatives such as those comprising the Vava'u development program require strong support from the local communities if they are to succeed in the long term. Intensification of agriculture and fisheries will only occur if the farmers and fishers wish those outcomes for themselves, and that, in turn, often requires their involvement in developing programs that tackle their specific concerns and needs. Moreover, the local knowledge they bring to the process will often improve the design of options and the subsequent evaluation and selection of the best option. This is especially the case where knowledge of the local environment can be very valuable in relation to projects such as road and causeway building. Local awareness of environmental conditions can help avoid basic problems arising from ill-considered locations or designs.

Second, environmental planning requires a good understanding of the environment in the area. It is not clear from the Phase II study that the social and natural environments were examined to establish parameters such as natural carrying capacities, resource use limitations, sustainable yields, limits of acceptable change, or other similar measures that are used to identify the basic environmental limits on economic and social activities. For instance, with water supply, it is not evident that a water resource survey was carried out to determine the extent of groundwater supplies and the potential sustainable production of water for consumption purposes. Expansion of tourism, crop processing, and general improvement of living standards will require extra water supplies. Is there a limit beyond which further usage would harm the groundwater resource and threaten social and economic activities? There does not appear to be an answer to this question in the report.

Third, and linked to the last point, the Atkins report was the stage at which an environmental assessment (EA) of development options would most usefully have been introduced. What are the environmental consequences of promoting more intensive agriculture in Vava'u? Will the expansion of tourism threaten the natural or social environment in some fundamental way? A SEA would have been very useful for exploring alternative development options and informing the policy process at that stage of the program.

Overall, it would seem that the Master Plan is based on earlier steps that explicitly recognize the importance of environmental management and the need to avoid adverse impacts on the social and biophysical environment. The specific topics to be covered in the Master Plan were generated within that context and it can be assumed that they were considered to be consistent with the underlying sustainability perspective of the Atkins report. However, there is no evidence of any formal examination of the possible environmental consequences prior to the selection of the particular development options. That does not mean those options are not the right ones for the Vava'u region and its people, but it does leave open the possibility that important environmental problems might still result from the initiatives being pursued.

Implications of Tourism Development

The Neiafu Master Plan includes an number of projects that directly or indirectly address tourism-related problems in the town. By undertaking these proposed projects, the tourism industry would be encouraged; in all likelihood, Neiafu would become a major center, attracting tourists in its own right and providing a service center function for tourists drawn to the wider region, especially the marine area. That encouragement could then result in greater tourism impacts in other parts of the region, particularly on the reefs and in popular anchorages. A project-oriented

EIA, assessing the immediate impacts of the proposed projects on the local environment, does not take this wider perspective into account, hence the need to assess the *policy of encouraging tourism* before developing specific projects for particular locations. The regional, and even national, implications need to be explored first, then development projects can be identified and planned within a framework that takes the wider implications into account.

Apart from the water supply question alluded to earlier, one particular issue that would need addressing in the light of the proposed developments is the question of solid waste management, especially domestic refuse. Population increase, allied with increasing standards of living, together with an increase in tourism activity, would inevitably lead to greater amounts of solid waste for disposal. This issue does not appear to have been considered in the development program to date, yet has serious environmental implications. At the time of the study solid waste was being dumped on a site immediately adjacent to a coastal mangrove area, approximately 2–3 km northeast of Neiafu. The location of the site was the worst possible, given the propensity of such sites to release toxic leachates into nearby water bodies, affecting water quality and the local aquatic biota. There did not appear to be any form of management operating at the existing site. Without a specific waste management strategy this situation would only get worse as the production of solid waste increased with economic development, producing long-term problems for the land and adjacent marine areas and for the health of the local people.

The biophysical impacts of tourism are reasonably well understood and recognized, but there is increasing concern around the world about the social impacts of tourism on small communities in areas of high natural values, such as Vava'u. The changes brought about in small communities can result in once-cohesive communities becoming disrupted, with alienation of young people, greater economic disparities between people involved in tourism and those still in traditional activities, increasing prices of goods for local people, commercialization of cultural activities, and other similar changes. Together with a degradation of the local biological and physical environment through tourist activities, affecting traditional food gathering areas or recreational and cultural areas, these changes can have dramatic effects on the local people.

It is important that the trade-offs between the undoubted benefits of developments such as tourism and the adverse effects on the social and biophysical environments are recognized and explicitly addressed by the local communities. Possibly the worst impact is a sense of helplessness as tourism growth seems to overwhelm a local community. Involvement in planning for tourism development, and retaining a strong interest in protecting the very resource base that supports the tourism industry, gives local people a sense of control that promotes a successful and sustainable tourism industry. This type and level of public involvement in the development of a sustainable tourism policy has clearly not taken place in Vava'u and one of the challenges facing societies such as this is to adapt traditional decision-making approaches to fill such gaps in the policy- and plan-making processes.

ASSESSMENT OF THE NEIAFU MASTER PLAN

As noted earlier, the Master Plan comprises a set of policies concerning land use and building standards, together with specific projects for upgrading infrastructure and enhancing the urban landscape in various ways. The scope of the Plan was largely determined by the terms of reference supplied by the Atkins report, but the relative emphases and general treatment of the issues in the Plan are a product of the planning study itself. This assessment looks briefly at the treatment used in the Plan, reinforcing some of the comments above, and then turns specifically to the environmental implications of the set of activities in the Plan.

EMPHASIS AND TREATMENT

It is immediately evident from the Plan that tourism is given a high degree of prominence. This is not altogether unexpected. The town has a central role to play in tourism services in Vava'u, and with a regional development policy that is seeking to enhance the primary sector activities as well as tourism, Neiafu's part in that policy will tend to reflect the urban-based services, and that tends to increase the emphasis on tourism issues.

However, the emphasis on tourism seems to go beyond what might be expected from the Terms of Reference (TOR). For instance, the proposals for creating a pedestrian-only area in the center of the town, allied with suggestions for the redevelopment of the buildings between that area and the proposed reclaimed water-front, indicate a strong tourism focus. Many of the infrastructure projects are presented in terms of their benefits, not only for local people, but also for tourism prospects.

In itself, this is understandable; the town does have the potential to be developed as a major tourist destination. But the broader policy basis of the Vava'u Development Program also emphasized the importance of improving the primary sector, as well as the basic living conditions for the local people. The emphasis of the Plan might, therefore, have been better placed on an appraisal of the role of Neiafu in the primary sector activities of the region, and then consideration of the projects could be made from the perspective of improving and enhancing that role. Wharf upgrading, market improvements, and road improvements, would then be seen clearly from that perspective. Pedestrian areas would be developed if local people thought this would be a real benefit to their quality of life, and/or it enhanced service functions. The Plan makes many proposals that may well improve the quality of life for the local people, but one is left asking the question: to what extent do the proposals match the concerns and needs of the local community? Who will benefit most from the proposals: the local people or the tourists?

The long term viability of the local economy and the local communities requires a balance of development intentions. The emphasis in the Plan does weigh strongly toward one particular economic sector, tourism. Moreover, there is no explicit recognition of the point made in the previous section, that encouraging Neiafu as a tourism center will increase tourism activity, and consequently, tourism impacts over a wider part of the region.

Project Assessment

A simple impact matrix was formulated to review the broad environmental implications of the various project proposals. The intention was not to provide definitive judgements about the environmental impacts of each project; properly constituted EIA studies would be necessary for such judgements. Instead, the matrix helped in the comparison of the broad environmental implications across the various projects and provided some guidance in identifying possible cumulative impacts and interactions between projects that might create enhanced impacts or other problems, such as hazards. It can also be used to examine the distributional aspects of impacts, that is, who is winning and who is losing from the overall set of projects.

There are major limitations to impact matrix analyses. Complex impact situations which involve spatially differentiated, direct, and indirect effects (perhaps also with temporal variation) are impossible to portray in a simple matrix. Hence, the matrix was used mainly as a device by the assessors to organize thoughts, and to portray in a simple fashion information about the possible impacts of the projects.

Main Points from the Impact Matrix

1) The assessment explicitly differentiated between the construction phase of the projects and their subsequent operational phase. It is clear that many, if not all, the proposed projects are likely to have adverse effects at the construction stage. These are mainly impacts such as noise, the disruption of daily social and economic activities, and the exposure of soil to rainfall and erosion processes during the construction period. The latter raises the real possibility of silt reaching the harbor during periods of heavy rainfall, affecting water quality, and perhaps also marine organisms, as well as aesthetic values.

2) Most of the adverse effects would fall on the local people and the marine environment, which is entirely as expected, given the projects will take place in a town by the sea. There are some indirect effects in the rural sector, mainly due to quarrying to provide crushed coral limestone for roads and pavement construction. The proposal to establish an environmental reserve would, by preserving a major area of comparatively unmodified forest, be the only project that would impinge directly on natural ecosystem components (soil, fauna, and flora). It can also be considered to have indirect benefits for the marine environment by not contributing silt (and sewage) as the nearby urban area does.

3) Local people should benefit from a number of the proposed projects in a variety of ways. The health of the local community, a key issue, would be improved by road construction (less silt, less airborne dust), better water supply (less chance of contamination), better sanitation (again, better groundwater quality), and better drainage (less overflowing of sewage tanks and pits, fewer puddles for insects to breed in, etc.). Social

activities, including daily living activities and interaction with other people, would be enhanced by many of the proposals, although adverse effects from construction activities will be felt (disrupted access to buildings or to certain parts of the town, dust and noise nuisance, etc.). Visual amenity in the urban area would benefit in the long term from most of the proposals, although all construction activities can be considered adverse amenity impacts (but unavoidable) in the short term.

4) Economic activities show an interesting pattern. Tourism would probably benefit in the long term from all the proposals. They all improve the look or the functioning of the urban area in a way that would be beneficial to some aspect of the tourist industry. Of the other sectors, agriculture and fishing would probably only benefit from the road and wharf construction projects (after short term inconveniences during construction). The service sector (including banks, shops, airline office, etc.) would benefit from some of the other projects, such as better drainage in the town center. Indirect benefits would also come from improvements in the agricultural and fishery sectors. However, the proposed removal of some old wooden buildings along the main street would obviously adversely affect the businesses currently located in them.

5) The last point indicates one important area of adverse impact suggested by the matrix — the historical and cultural aspects of the urban landscape. Loss of old buildings and the loss of a central park would be changes that local people might not wish to see.

COMMENT

It seems likely that most of the impacts of the various activities proposed in the plan would be social impacts. The effects on people, both beneficial and adverse, would probably be more important than the likely effects on the natural environment. Even the effects on the marine environment can be seen in many ways as effects on the local community, by limiting fishing, recreation, amenity, and tourism use of the marine area.

When considering the likely impacts, we have assigned the labels "adverse" and "beneficial" depending on how we think various sectors of the community might respond to the changes. But these are essentially value judgements, and ultimately the local people must make the judgement about the changes that might occur as a consequence of the proposals. They must decide if the old buildings along the main street are cultural assets worth saving or dilapidated buildings that ought to be removed for site redevelopment. This type of value decision involves trade-offs of benefits and costs and it is the local community that has to live with the results of the trade-off calculations. For this reason it is important that the people have some involvement in considering the proposals and their likely effects.

In this regard one might ask the local community if they want the benefits of improved visual and amenity values in the town center, and enhanced tourism values, especially when some of the local people will have to bear the cost of busier roads, with associated dirtier, noisier, and more hazardous living conditions, and especially when the agricultural and fishing sectors will primarily benefit from just two of the projects. There has not been public involvement of this kind, and, as implied earlier, the Plan seems to be planning *for* people, rather than planning *with* people.

In terms of cumulative impacts, the most obvious one is the combined effects of the construction phases of all or several projects if they were to be undertaken concurrently. Being in a comparatively small geographical area, the construction impacts could be severe: disruption of traffic and pedestrian access, noise, dust, and the danger of large quantities of silt entering the harbor. Although the timetable for implementing projects would be determined largely by the availability of funds, some thought would have to be given to the possibility of coordinating the construction phases of certain projects to minimize the potential for social and economic disruption and silt transport.

Not surprisingly, the marine environment would be the major recipient of many of the possible adverse effects of the proposals, largely through the silt problem. However, the combined benefits would be large if the proposals were effective: reduced sewage from seepage, surface run-off, and yachts, and reduced silting in the long term as a result of sealed roads. At the same time, it must be recognized that a substantial contribution to the silt problem comes from the activity of pigs in local gardens, removing ground vegetation over large areas, and exposing surface soil to erosion by rainfall. A comprehensive approach to managing marine water quality and the sedimentation problem in the harbor area would have to include strategies for dealing with this issue. One way to tackle this is to involve the community in recognizing and solving the problems with the marine environment; solutions to pig management (greater use of pens, for instance) would need to be initiated by the community and enforced by social pressure and sanctions if they were to be successful.

Impact Assessment Needs

Several proposals would probably need closer scrutiny for their environmental implications through project-level EIA. The main candidates for further work include the following:

a) The proposal to develop a sewerage system for central Neiafu, collecting the sewage generated in the area and piping it out to sea, would need serious scrutiny, with the various options being considered in terms of their economic, technical, and especially, environmental feasibility. Sewage is probably the most serious threat to the marine-based economy (fishing and tourism) and any such proposal must be rigorously appraised.

b) A social impact assessment might be considered for the general set of proposals that would alter the physical nature of the town, including the change in the location of the market, the pedestrianization of the town center, and the re-routing of the main road around the center. The nature and extent of the cumulative effects of these proposals might not be anticipated by the local community and a social impact assessment would be a useful vehicle to address any concerns that might be evident before the proposals go any further.

c) While not strictly an impact assessment, a study of the local waste management system is an urgent requirement. The development proposals, if they have the desired effect, will increase the level of economic activity in the town, and probably result in a steady population increase over the next few years. Domestic refuse disposal must be managed more effectively than it appears to be at present, if environmental and amenity problems are to be avoided.

MONITORING REQUIREMENTS

Neither the Atkins report nor the Neiafu Master Plan makes provision for environmental monitoring activities; consequently, there would be no mechanisms for detecting adverse changes in the environment and formulating management responses. This is an important issue as economic activity in the primary sectors and in tourism increases, and it will only be sustainable if it does not begin to damage the resources on which it is based. Monitoring is the only way this can be detected, which is the first step in developing a management response.

Monitoring priorities in relation to the Neiafu proposals would be:

1) the marine environment in Port of Refuge and around the outer islands: water quality (turbidity, faecal coliforms, etc.); indicator organisms (algae; benthic organisms; diversity of reef fish, etc.); habitat indicators (health and integrity of coral, etc.); sedimentation processes adjacent to the town (and other local population centers); etc.
2) the Neiafu community, and the wider population of Vava'u: age/sex make-up of the population; household structure; employment characteristics; health indicators; public order indicators; attitudinal indicators; etc.

SUMMARY AND CONCLUSIONS

MAIN ISSUES

This SEA identified a number of issues:

a) There is a basic sustainable development objective evident in the Vava'u Development Program (VDP), especially as expressed through the Atkins report. However, there does not seem to have been any formal EA of development options during the early stages of the Program. Nor did there appear to have been any studies into such parameters as carrying capacities, environmental constraints, etc., that would indicate an environmental planning approach to the development program.
b) The Master Plan seems to accord tourism development a higher priority than that implied in the Atkins report. At the same time there is no apparent awareness of the wider, regional, and environmental implications of upgrading tourism facilities in Neiafu.

c) Most of the effects of the various proposed projects will probably fall on the local community and the marine environment. The construction phase of several projects would probably lead to silt transport to the marine environment, especially near the harbor area. Social and amenity impacts tend to be adverse in the short term and positive in the longer term; health, in particular, is likely to improve if the proposals are implemented. The combined social impacts of the various proposals affecting the town, itself, could be quite marked.

d) Agricultural and fishing activities would probably benefit mainly from two proposals (road upgrading and wharf redevelopment). Tourism would benefit from all the proposals.

e) Monitoring would be required for the marine environment, coming under pressure from tourism as well as fishing, and for the local community of Neiafu, to allow for early identification of social problems resulting from the projects.

On a more general note, the extent of community participation in the development and planning process might be reviewed. It is difficult to draw definite conclusions from the reports examined in the course of the study, but it would appear that the local communities have not been given many opportunities to contribute their ideas or make known their concerns and values. Greater public participation, perhaps based on traditional local practices, would help in the design of more appropriate and effective development proposals, and provide stronger community support for the proposals.

Study Conclusions

In conclusion, the environmental consequences of the proposed development projects for Neiafu have not been closely examined in the Master Plan, nor have the wider environmental implications been considered in the earlier Atkins report. The nature of the proposals, enhancing existing activities, means that there are unlikely to be major environmental problems arising in the short term. However, the increasing and cumulative effects of tourism are important, and it might be advisable to initiate an environmental management program targeted at the tourism sector. This would examine the current extent of tourism and the impacts (both beneficial and adverse) experienced to date. But it would also be advisable to consider the environmental factors that would limit the expansion of tourism in the region (environmental sensitivities; natural hazards; natural carrying capacities, etc.). In this way, the very features that attract visitors can be managed and protected in the longer term to sustain the economic viability of the sector.

Benefits of the SEA

This SEA was an evaluation of an existing plan, and to that extent, it differs from other applications of SEA. The benefits of the assessment are manifested in two ways. First, SEA used in this way provides a means for evaluating the policy and plan development process, itself, from the perspectives of sustainable development and environmental protection. Hence, it can be viewed as an *environmental policy analysis* tool. Second, and more typically, the SEA addresses the environmental implications of the plan, and suggests issues that will need to be considered more fully if the plan, or parts of it, proceed.

1) *SEA and the policy/plan development process.* The case study demonstrated the lack of attention in the policy and plan development process to various issues that we now associate with sustainable development. These include:
   - Explicit reference to the carrying capacity of the local environment, and specific systems within it. For example, there was no obvious investigation of the water resources of the main island, nor was the regional environment assessed for its ability to sustain tourism without significant degradation.
   - The overall quality of life of the local people. The early reports clearly saw economic improvement of the local population as a priority, but there was little attention paid to other aspects of community life, especially social and cultural considerations. Consequently, potential impacts on these aspects of local life were not accorded sufficient importance in the development of the plan.

   The SEA demonstrates that the policy and plan development processes in Tonga, and possibly in other Pacific Island nations, might need to be reviewed if truly sustainable resource development is to be achieved.

2) *SEA and environmental implications of the plan.* The assessment placed particular emphasis on the interaction of the various activities proposed under the plan and their likely cumulative impacts, on the local people and the biophysical environment. This suggested new issues to be considered, and helped the assessors identify gaps in the plan, such as the lack of a waste management strategy. It is also clear that closer environmental scrutiny is required for various components of the plan, with project-level impact assessments indicated for activities such as the wharf development, market relocation, and sewerage infrastructure upgrade. This reinforces the tiered nature of impact assessment, with strategic level assessments being used to identify where project-level assessments are required, and preparing the ground for them.

The actual impact of the SEA on decision-making in this particular case is hard to gauge, as the assessment was carried out quite late in the planning process. Since the Master Plan was released, the VDC and the VDU have increasingly focused on the key infrastructure improvements contained in the Plan. The priorities are now seen to be road improvement, the upgrading of the wharf, and the improvement of the town water supply. The latter is being addressed by the Tongan Water Board as part of its national program of water supply improvement. The other two projects are being tackled by the VDU.

In effect, the main tourism related proposals, such as the foreshore redevelopment, and the construction of new retail outlets to cater to tourists, have been set aside in favor of projects that will improve the immediate quality of life of the local people and enhance economic activities in the agricultural, fishing, industrial, and service sectors. Overall, the VDC seems to have reached similar conclusions as this SEA study, that the Master Plan leaned too strongly toward the enhancement of the tourism potential of the town, and that the local people would benefit more from the infrastructure projects rather than the beautification projects.

Economic and political considerations probably combined to produce this outcome. However, the SEA provided strong environmental (including social) grounds for caution, and demonstrated that a similar assessment, carried out earlier in the process would be of benefit in identifying potential environmental and resource development problems. The true worth of the exercise will only be realized when SEA is used by planners and environmental managers in Tonga and other Pacific Island nations as a natural tool for protecting the environment and promoting sustainable development.

ACKNOWLEDGMENTS

The original study, on which this chapter is based, was carried out for SPREP and is fully reported in publication number 98 in the *SPREP Reports and Studies Series*. The opinions expressed in this chapter are entirely those of the authors and do not in any way represent the views of SPREP.

REFERENCES

1. Theroux, P., *The Happy Isles of Oceania: Paddling the Pacific*, Penguin, London, 1992, 434.
2. *Neiafu Master Plan, Draft Final Report*, Kinhill Kacimaiwai, in association with Kinhill Riedel & Byrne, 1993.
3. *Vava'u Regional Development Program, Phase II: Final Report, Project and Program Dossiers*, WS Atkins International, Epsom, England, 1989.
4. *Neiafu Urban Development Project, Technical Feasibility Study*, Kinhill Kramer Pty Ltd., Boroko, PNG, 1994.
5. *6th Development Plan 1991–95*, Central Planning Unit, Government of Tonga, 1990.
6. *Environmental Management Plan for the Kingdom of Tonga*, ESCAP, Bangkok, 1990.

15 SEA of Water Management Plans and Programs: Lessons from California

Ronald Bass and Albert Herson

CONTENTS

INTRODUCTION

In California, water management is an extremely complex public policy issue involving the interplay of many environmental, economic, legal, and political factors. With an arid climate, diverse physical environment, shortage of public funding, and rapidly increasing population, the storage, delivery, and usage of water is one of the state's most critical planning issues. Consequently, strategic, long-range planning is a high priority among many of the state's hundreds of water management agencies. Each year, many of these agencies prepare Program Environmental Impact Reports (EIR) (the California terminology for strategic environmental assessments [SEA]) that evaluate the environmental impacts of water management plans and programs. These Program EIRs often involve the comparison of strategic alternatives and contain innovative approaches to environmental impact evaluation and mitigation of significant effects. This chapter:

- summarizes some of the key issues involved in water resource planning in California
- explains the interrelationship of environmental issues with politics, economics, and geography within the state as they relate to water
- identifies the key water management agencies that typically prepare Program EIRs
- describes the types of water-related activities for which Program EIRs have been prepared
- summarizes how environmental impacts are typically evaluated in long-range water management programs
- introduces recommendations that other nations may rely on to prepare SEAs on water resource programs

FRAMEWORK FOR WATER MANAGEMENT IN CALIFORNIA

CALIFORNIA'S UNIQUE GEOGRAPHIC SITUATION

To understand the importance of programmatic environmental impact reports in California, a brief explanation of the state's unique geographic and socio-economic factors is necessary. The relationship between natural conditions, population patterns, and social and economic factors strongly influence water management in the state. California's climate is characterized by distinct wet and dry seasons and by great variations in precipitation between the northern and southern parts of the state. More than 75% of the state's rainfall occurs in the north, and more than 75% of the state's population lives in the south. Additionally, most of the state's major cities are found along the Pacific coast, while most of the state's snowpack (which contributes greatly to the water supply) is in the Sierra Nevada mountains along the eastern part of the state. Consequently, most of the state's largely urban population of more than 32 million persons use water that originates from somewhere else. This trend is expected to continue as the state's population increases to a projected level of approximately 47 million by 2020.[1]

In addition to domestic usage of water by the state's large and growing population, agriculture represents a substantial share of total water usage. According to the California Department of Water Resources, agriculture accounts for 43% of water usage, whereas urban uses only account for 11%. Water is a critical element in California's agricultural economy, which is one of the most diverse and important in the world. Like urban water supplies, most of the state's agricultural water users rely on water that originates in other parts of the state.[2]

Besides the state's surface water system, in some parts of California groundwater is relied on by both urban and agricultural users. However, many of the state's groundwater basins are seriously depleted, resulting in poor quality water for users and subsidence of the land above. The quality and quantity of groundwater resources are continuing management concerns of the state's water management agencies.

Throughout the history of California, the competition for water between urban and agricultural users has been intense because these users often rely on water from the same sources. To complicate this conflict, California has had neither a comprehensive, nor consistent approach for allocating water nor for delivering water to users.[3]

The longstanding conflicts between urban and agricultural water users have been complicated further in recent years as the demands for "in-stream" uses of water, to sustain fish and wildlife, have escalated. The protection of an adequate supply of in-stream water has been high on the political agenda of many environmental organizations, as well as a goal of some of California's natural resource protection agencies. The listing of many aquatic species as "endangered" or "threatened" under the federal and state Endangered Species Acts has magnified the three-way conflict between water users because a reliable water supply is critical to the survival of many listed species. Currently, approximately 46% of water usage goes to environmental protection.

As a result of these unusual geographic, demographic, and economic factors, government agencies have developed a complex water management system to move water from one part of the state to another. Federal and state agencies, as well as local water districts, have developed and operate a massive system for water storage and delivery that crisscrosses California (see Figure 15.1). An equally complex legal system has developed by which water rights are established, bought, and sold among the state's hundreds of water agencies.

To adequately prepare for ever-increasing future water needs, California's water management agencies and districts are constantly developing new plans and programs for providing additional water supply and distribution. Thus, each year dozens of programmatic environmental impact studies are prepared on a broad variety of water plans and programs.

GOVERNMENT ROLE IN WATER MANAGEMENT

Water planning and management in California is the responsibility of many federal, state, regional, and local agencies, as well as hundreds of water districts.[4] Table 15.1 summarizes the roles of the major agencies involved in water management. All the federal agencies listed in Table 15.1 are subject to the National Environmental Policy Act (NEPA) and must prepare environmental impact assessments (EIA) as required by NEPA[5] (see Chapter 2). All the state and local agencies and water districts in California are subject to the California Environmental Quality Act (CEQA) and must prepare EIA documents under that law. For some water projects, both federal and state agencies are involved, thus triggering the requirements of both laws. The focus

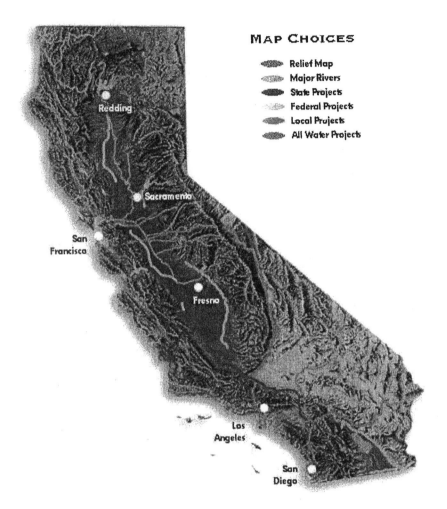

FIGURE 15.1 California major water projects. (Available at: http://www.dwr.water.ca.gov/dir-CA_mapsR2/All_Projects_MapR3.html. With permission.)

of this chapter, however, is on CEQA and its application by state and local water management agencies.

PROGRAMMATIC ENVIRONMENTAL ASSESSMENTS IN CALIFORNIA

APPLICABILITY OF CEQA TO WATER MANAGEMENT PLANS AND PROGRAMS

CEQA applies to a broad range of government activities. It applies to water management plans and programs only when the water management agency makes a dicretionary decision that has the potential to significantly affect the environment.

TABLE 15.1
Key Agencies Involved in Water Management in California

	Agency	Responsibility
Federal Agencies	U.S. Bureau of Reclamation [http://mp.usbr.gov/]	The Bureau manages, develops, and protects water and water-related resources in an environmentally and economically sound manner for the American people. In California, one of the Bureau's main functions is the operation of the Central Valley Project, a system of major dams and canals that provides water to vast areas of the state.
	U.S. Army Corps of Engineers [http://www.usace.army.mil/]	The Corps of Engineers is responsible for maintaining the quality of the nation's waterways. The Corps regulates all activities involving dredging and filling in waterways, including the nation's wetlands.
	U.S. Fish and Wildlife Agency [http://www.fws.gov/index.html]	The U.S. Fish and Wildlife Agency maintains more than 500 wildlife refuges throughout the country. Additionally, the agency is responsible for implementing the federal Endangered Species Act.
	National Marine Fisheries Service [http://www.nmfs.gov]	The National Marine Fisheries Service protects marine resources, including endangered species, under the federal Endangered Species Act.
	U.S. Environmental Protection Agency [http://www.epa.gov/]	The Environmental Protection Agency oversees state implementation of the Clean Water Act and the regulation of wetlands.
State Agencies	California Department of Water Resources [http://wwwdwr.water.ca.gov/]	The California Department of Water Resources (DWR) develops, manages, and delivers water resources to users throughout California. One of DWR's main functions is the operation of the State Water Project, a system of major dams and canals.
	State Water Resources Control Board [http://www.swrcb.ca.gov]	The mission of the State Board is to ensure the highest reasonable quality of waters of the state, while allocating those waters to achieve the optimum balance of beneficial uses. The State Board implements the federal Clean Water Act and adjudicates water rights between competing users.

TABLE 15.1 (CONTINUED)
Key Agencies Involved in Water Management in California

Agency	Responsibility
Regional Water Quality Control Boards [http://www.swrcb.ca.gov]	Nine Regional Water Quality Control Boards implement the federal Clean Water Act and similar state laws throughout in various regions of the state.
California Department of Fish and Game [http:// www.dfg.ca.gov]	The mission of the Department of Fish and Game (DFG) is to manage the state's fish, wildlife, and plant resources, and their habitats, for their ecological values and their use and enjoyment of the public. DFG also implements the California Endangered Species Act.
Water Districts Hundreds of water districts have been formed within California, each covering different geographic areas. Some of the major water districts include:	Each water district is established under a separate law and is typically responsible for supplying water to a particular geographic area. Most water districts develop long-range master plans and programs that are subject to the requirements of CEQA.
• Metropolitan Water District of Southern California [http://www.mwd.dst.ca.us]	It provides water to much of the Los Angeles region.
• San Diego County Water Authority [http://www.sdcwa.org]	It provides water to much of the San Diego region.
• East Bay Municipal Utility District [http://www.ebmud.com]	It provides water to many of the cities in the San Francisco Bay Area.
Cities and counties Many of California's 558 cities and 58 counties are engaged in some level of water planning and management [http://www.ceres.ca.gov]. Additionally, some cities have developed and operate their own water systems. Two of the larger of such systems are:	Cities and counties are often the final provider of water to their citizens and often develop water plans and programs.
City and county of San Francisco Water Department	It provides water to the City of San Francisco
Los Angeles Department of Water and Power [http://www.ladwp.com]	It provides water to the City of Los Angeles.

The detailed procedures for screening, scoping, and preparation of EIA documents are described in the State CEQA Guidelines issued by the California Resources Agency. According to the State CEQA Guidelines, activities that are subject to CEQA include policies, plans, and programs (PPP) [40 Cal. Code Reg. 15378]. Thus, many of the policy-making and planning activities of water management agencies are subject to the requirements of CEQA. Unlike EIA laws in other nations, CEQA has always applied to such higher-level decision-making. As a result of this requirement, agencies throughout California routinely prepare Program EIRs. For example, in 1996, of the 401 EIRs submitted for review by the state clearing-house, 101, or 25%, were prepared for PPP.

Under CEQA, when a water management agency commences a planning process or program, it must first conduct a two-level screening process. The first screening is necessary to determine if the activity in question is subject to an exemption. The second screening, known as an "Initial Study," is necessary to determine if the proposed plan or policy may have a significant effect on the environment. If potentially significant environmental impacts are likely to occur from future activities under the plan, the agency must prepare an EIR. If, however, the agency determines that the proposed plan or program would not result in significant environmental impacts, it may prepare a Negative Declaration, which is an abbreviated assessment of the environmental implications. The decision as to which level of CEQA review is necessary lies with the agency proposing the plan or program. Most water management plans and programs require EIRs because they typically cover broad geographic areas, lead to future growth, and usually result in a variety of significant environmental effects. According to the State CEQA Guidelines, the procedures for preparing EIRs for plans and programs and the required contents of those documents, are essentially the same as for those of project-level EIRs.[6]

Despite the procedural similarities between the two types of EIRs, the State CEQA Guidelines provide some special guidance for preparing EIRs for plans and programs. The document typically prepared for these strategic activities is known as a "Program EIR." Program EIRs generally analyze the broad environmental effects of the plan or program with the acknowledgment that site-specific environmental review may be required for individual components of the plan or program. Program EIRs are usually prepared for activities that are linked geographically, are logical parts of a chain of contemplated events, or are carried out under the same authorizing statutory or regulatory authority and have similar environmental effects. In keeping with this guidance, water management plans and programs are prepared as Program EIRs.

The State CEQA Guidelines further cite five specific advantages of Program EIRs that distinguish them from Project EIRs. Specifically, Program EIRs:

- provide an occasion for a more exhaustive consideration of effects and alternatives than would be practical in an EIR on an individual action
- ensure consideration of cumulative impacts that might be slighted by a case-by-case analysis
- avoid duplication and unnecessary reconsideration of basic policy considerations
- allow agencies to consider broad policy alternatives and program-wide mitigation measures at an early time when the agency has greater flexibility to deal with such issues
- allow agencies to reduce paperwork through "tiering"
- if a Program EIR is prepared in sufficient detail, it may be used not only to evaluate the impacts of the plan or program, but also to make individual

project decisions. Sometimes, however, a Program EIR is quite general, and additional project-level EIRs must be prepared to implement the program. This situation, known as "tiering," is discussed below.

TIERING

EIRs prepared for water management plans and programs often serve as the first tier of a multi-level planning process. According to the State CEQA Guidelines, "tiering" refers to the concept of a multi-level approach to preparing EIRs. A first-tier EIR typically covers issues in a broad, generalized level of analysis.[7] When individual projects are proposed under the plan or program, second-tier EIRs are typically prepared for that focus on the project-specific impacts. Tiering is a method to streamline EIR preparation by allowing agencies to focus on the issues that are ripe for decision, and exclude from consideration those issues that have already been decided or are not yet ready for decisions. In California, agencies are encouraged to tier their environmental documents whenever possible.

TYPES OF WATER MANAGEMENT PLANS AND PROGRAMS SUBJECT TO CEQA

Because of the diverse range of issues facing water management agencies in California, some of these agencies prepare Program EIRs yearly on a broad variety of plans and programs, including:

- state water agency management programs
- water district master plans
- water transfer agreements
- flood control plans
- river basin plans
- water reclamation and reuse plans
- groundwater management plans
- water quality improvement programs
- city and county water elements of comprehensive plans

A list of the EIRs prepared for water plans and programs in California between 1997 and 1998 is shown in Table 15.2.

TYPICAL ISSUES EVALUATED AND METHODS OF IMPACT ASSESSMENT

Program EIRs prepared for water plans and programs typically evaluate the impacts of a range of strategic alternatives in comparative form. Although no two EIRs contain exactly the same issues or methods of analysis, some common recurring topics are covered in most documents. In California, many agencies have learned that it is possible to quantify the environmental impacts of plans and programs. Thus, many of the same methods used for project-level analysis are also used at the plan or program level. However, the analysis is often more generalized and less site-specific than at the project level. For example, the basic

TABLE 15.2
Environmental Impact Reports Prepared for Water Plans and Programs in California 1997–1998

Title	Lead Agency/Year	Type of Activity	Summary
Alameda County Water District	Integrated Resources Plan (1998)	water supply	capital Improvement program for future water facilities
Monterey County Water Resources Agency	Salinas River Vegetation Management Program (1998)	flood management	alternative approaches to vegetation management in river corridor
Belridge Water Storage District	Transfer of State Water Entitlement (1998)	water transfer	transfer of water from the State Water Project from three water districts to other users
California Resources Agency (joint state/federal effort)	CALFED Bay-Delta Program (1998)	water management plan	long-term plan to restore ecosystem and improve water management in the San Francisco Bay–Sacramento/San Joaquin River Delta
City of Chino Hills	Chino Hills Master Plan of Water Supply & Distribution (1998)	water supply	master plans for water supply and distribution
Stanislaus County	Diablo Grande Specific Plan Water Resources Plan	water supply	long-term water supply for a 29,000-acre resort development
Alameda County Flood Control Zone 7	Berrenda Mesa Water District	water transfer	transfer of water between water districts
City of Escondido	Regional Water Reclamation & Disposal Program (1997)	waste water reclamation and disposal	study of three alternatives for increasing disposal capacity and water re-clamation
Sacramento County Water Agency	Central Valley Project Water Supply Contracts (1997)	water supply	sale of water rights from the federal Central Valley project to local water district
City of Napa	Water System Optimization and Master Plan (1997)	water supply	program to improve water system and change service area boundaries
San Luis Obispo County	Nacimiento Water Supply Program (1997)	water supply	program to deliver water from Lake Nacimiento to 18 water purveyors
Lake Arrowhead Community Services District	Water & Wastewater System Improvements (1997)	water supply and disposal	update water and waste-water master plans

TABLE 15.2 (CONTINUED)
Environmental Impact Reports Prepared for Water Plans and Programs in California 1997–1998

Title	Lead Agency/Year	Type of Activity	Summary
Rancho California Water District	Water & Wastewater Plan (1997)	water supply and disposal	update water and waste-water master plans
Santa Clara Valley Water District	Upper Guadalupe River Flood Control Project (1997)	flood control	program for improving flood management
California Department of Water Resources	State Water Project Supplemental Water Purchase (1997)	water transfer	water transfers between State Water Project and various water contractors

Source: State of California, Resources Agency[8]

TABLE 15.3
Typical Issues and Methodologies of Impact Analysis in Programmatic Environmental Impact Reports

Environmental Issue	Typical Methodology
water quality	water quality modeling, sampling
hydrology	hydrological modeling
fisheries	field surveys, statistical sampling, ecological modeling, consultation with federal and state fisheries agencies, quantification of species change
vegetation and wildlife	aerial photography, field surveys, ecological modeling, consultation with wildlife agencies, quantification of habitat changes
socio-economic impacts	quantitative projections of program costs and economic effects on water users
growth inducement	quantitative projections of like new development supported by additional water supplies
land use impacts	quantification of anticipated shifts in uses (e.g., agricultural to urban)
soils and geological impacts	mapping, quantification of sedimentation
cultural resources (e.g., historic and archeological)	mapping, consultation with federal and state authorities
economics	quantification of project costs and economic impacts

assumptions on which computer models are made are often based on forecasts and assumptions about future conditions. Additionally, Program EIRs are generally comprehensive and interdisciplinary and often rely on quantitative methods of forecasting future impacts. Table 15.3 summarizes some of the typical issues and methodologies of impact analysis found in EIRs for water plans and programs.

Case Study — The East Bay Municipal Utility District Water Supply Program Environmental Impact Report

A representative example of a Program EIR for a water management plan is the *Water Supply Management Program* for the East Bay Municipal Utility District (EBMUD) near San Francisco (refer to Figure 15.2). This programmatic EIR is summarized in Box 15.1.

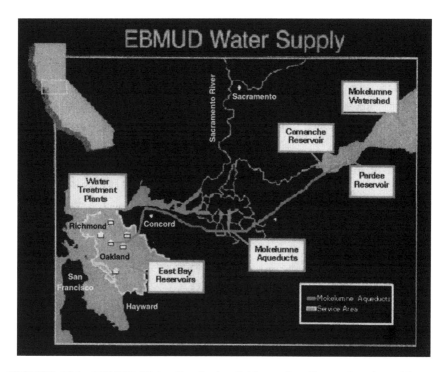

FIGURE 15.2 EBMUD Water Supply (available at: http://www.ebmud.com/ebmwater/wsupply.html).

Box 15.1 – EBMUD Water Supply Management Program
Environmental Impact Report[9]

Summary of the EIR:

    EBMUD is a publicly owned water district that serves water to 1.2 million customers and provides wastewater treatment for 600,000 customers residing in portions of Alameda and Contra Costa Counties in California (east of San Francisco). The EBMUD Water Supply Program Environmental Impact Report evaluated the environmental impacts of several strategic alternatives for providing water to a large metropolitan community for 25 years in the future.
Objectives of the program were to:

- Meet the increasing water supply demand of future customers

---

- Replace reductions to existing supply due to increased usage by nondistrict users
- Replace reductions to existing supply due to in-stream use allocations (e.g., fish)
  - Avoid possible water shortages from droughts

Strategic Alternatives Evaluated:

1. Demand-side management – Reduce demand through water conservation.
2. Groundwater sources – Increase pumping of groundwater.
3. Delta Supply – Obtain new water supplies from federal and state agencies that control water in the Sacramento River/San Joaquin River Delta.
4. Folsom South Canal Connection – Obtain additional supplies from the U.S. Bureau of Reclamation by developing a canal from Folsom Rese.
5. Raise Pardee Dam – Raise the dam and thereby increase capacity at the district's existing major reservoir east of Sacramento.
6. No Action – Take no new action and allow demand to exceed supply in the future.

Impact topics where there were important distinctions between the alternatives:

- economics
- growth inducement
- recreation
- water quality
- wetlands and riparian habitats
- fisheries

Public Involvement in EIR preparation:

- citizens advisory committees representing non-governmental organizations (NGOs) and affected communities
- regular mailings to all interested parties
- issue-oriented public meetings
- inter-agency forums
- board of director's workshop
- public review of Draft EIR (680 written and oral comments received)

Results:

- adoption of the Folsom South Canal Alternative with some groundwater supply
- development of a detailed database concerning affected environmental resources
- preparation of "second-tier" EIR is underway on the selected alternative for the construction of the Folsom South Canal connection

PROGRAM EIRS HELP ACHIEVE CEQA'S OBJECTIVES

Program EIRs have become an integral part of water planning in California. Many of California's key water management agencies have considerable experience preparing EIA for their plans and programs. Although CEQA requires Program EIRs to adhere to specific procedural requirements, it affords water management agencies considerable flexibility as to how they integrate environmental factors into their plans and programs.

The integration of EA with water plans and programs has helped water management agencies achieve the objectives of CEQA. Some of these results are summarized below.

ENCOURAGING INFORMED DECISION-MAKING

The integration between EIA and water management agency planning and programming has greatly increased the extent to which environmental factors are considered at the strategic level of decision-making. As a result of CEQA's requirements, water management agencies typically consider environmental issues at all stages of planning and plan implementation.

MITIGATING ENVIRONMENTAL IMPACTS

The consideration of environmentally superior alternatives and mitigation measures has become common practice under CEQA. Although a water management agency is not required to select the environmentally best alternative, even if they do not, they must adopt mitigation measures when they are feasible. One of the main advantages of preparing a Program EIR is that it can be used to develop baseline environmental information and generic mitigation measures that can then be applied on a project-by-project basis as individual projects are proposed under the plan or program. Many water management agencies use this approach to streamline the environmental review process at the project-level.

ENHANCING PUBLIC INVOLVEMENT IN THE PLANNING PROCESS

The application of CEQA to water agency plans and programs has greatly enhanced the opportunities for the public to participate in water policy in California. CEQA's procedures for preparing Program EIRs include the same opportunities for public involvement that are typically found in project-level review, such as scoping, public notices, opportunities to comment, and monitoring. In many Program EIRs, hundreds of individuals and organizations often participate. Thus, the public viewpoint is routinely considered in water decisions.

FOSTERING BETTER INTERGOVERNMENTAL RELATIONS

The implementation of environmental assessment (EA) at the plan and program level has fostered better intergovernment relations between state, federal, and local agencies and districts. It has created new opportunities for agencies to comment on, and thereby influence, the outcome of other agency's planning. Additionally, as a result of the EIA requirements of CEQA and NEPA, many water management agencies are now developing cooperative approaches to planning. For example, in one of the most ambitious large-scale programs, federal and state agencies are cooperating on a major planning effort to manage water in the Sacramento/San Joaquin River Delta. In furtherance of this effort, the agencies recently released a joint federal-state EA.[10]

STREAMLINING PROJECT LEVEL DECISION-MAKING

As the EBMUD case study illustrates, tiering is a useful method for preparing EIAs of water plans and programs. Most Program EIRs are eventually used as first-tier documents when project-level environmental review is conducted. This multi-tiered approach to planning has greatly streamlined the environmental review process for many agency decisions.

CONCLUSION — LESSONS FROM CALIFORNIA

As other nations consider the adoption of SEA requirements, they should familiarize themselves with the California experience with water management plans and programs. If they do, they will find very few disadvantages to applying EA to plans and programs. Although Program EIRs often result in some additional cost and time, offsetting savings can often be achieved at the project level. Further, the preparation of program EIRs also enables agencies to realize additional, often unquantifiable, benefits such as increased public involvement, improved intergovernmental relations, and more informed decision-making, all of which will hopefully result in better strategic decisions.

REFERENCES

1. California Department of Finance, Interian County Population Projections, Demographic Research Unit Sacramento, California, 1998. On-line: http://www.dof.ca.gov/htm/Demograp/p1netar.htm.
2. California Department of Water Resources, *California Water Plan Update – Public Review Draft*, The California Resources Agency, Sacramento, California, 1998.
3. Office of Planning and Research, *California Water Atlas*, Sacramento, California, 1979.
4. California Resources Agency, *List of State and Federal Environmental Agencies in California*, Environmental Resources Evaluation System, Sacramento, California, 1998. On-line: http://www.ceres.ca.gov/.
5. Bass, R. and A. Herson, *Mastering NEPA: A Step-by-Step Approach*, Solano Press Books, Point Arena, California, 1993.
6. Bass, R. and A. Herson, *CEQA DESKBOOK: A Step-by-Step Guide on How to Comply with the California Environmental Quality Act,* 2nd edition, Solano Press Books, Point Arena, California, 1999.
7. 40 California, Code of Regulations, Section 15385.

8. California Resources Agency, *State Clearinghouse Environmental Impact Assessment Data Base*, 1997-1998 in the California Environmental Resources Evaluation System, Sacramento, California, 1998. On-line: http://www.ceres.ca.gov/ceqa.

9. East Bay Municipal Utility District, *Updated Water Supply Management Program Final Environmental Impact Report*, Oakland, California, 1993.

10. CALFED Bay Delta Program, *Executive Summary Programmatic Environmental Impact Statement/Environmental Impact Report on the CALFED Bay Delta Program*, California Resources Agency/U.S. Federal Bureau of Reclamation, Sacramento, California, 1998. On-line: http://calfed.ca.gov/current/executive_summary/html.

16 SEA in the Selection of Projects to Restore Louisiana's Coast

Lee Wilson and Anna Hamilton

CONTENTS

INTRODUCTION

The Mississippi River has created one of the largest delta systems in the world. Its associated wetlands include more than 1.6 million hectares (4 million acres) of fresh, intermediate, brackish, and saline marsh, as well as cypress-tupelo swamp and bottomland hardwood forest. These wetland habitats are an essential component of an estuarine ecosystem that supports one of the world's most productive fisheries.

In the natural system, the building of new habitat in the active delta was approximately offset by the natural loss of land due to subsidence and erosion in abandoned delta lobes. Human development has caused wetlands loss rates to increase dramatically, to at least 65 km² per year (25 mi² per year), or an average of about 0.4% per year. Factors causing loss include reduced flooding and sediment from the Mississippi River because of flood control levees and upstream dams, and extensive dredging of navigation and oil and gas canals. The altered hydrology has allowed increased saltwater intrusion and reduced vegetative production, so that many marshes no longer keep up with natural subsidence.

This problem is considered one of national importance in the U.S., because the wetland losses in coastal Louisiana represent about 80% of all the coastal wetland losses occurring in the continental U.S., because the wetlands represent a unique ecosystem of intrinsic value; and because these wetlands support an impressive list

1-56670-360-3/00/$0.00+$.50
© 2000 by CRC Press LLC

of associated natural resources. The commercial fisheries alone represent about 25% of the finfish landings in the U.S. and about 40% of its shellfish landings.

The U.S. Congress has authorized the spending of about $40 million per year for restoration (loss-prevention) projects, such as shoreline protection, hydrologic restoration, river diversion, and marsh creation with dredged material. The funding is provided through the Coastal Wetlands Planning, Protection, and Restoration Act (CWPPRA).

The process for allocating CWPPRA funds represents strategic decision-making at the programmatic level for the specific purpose of maximizing environmental benefits. Thus, the funding process requires a type of Strategic Environmental Assessment (SEA) to select those projects with the best environmental performance. This use of SEA is in addition to the project-specific impact analyses that are done in accordance with the National Environmental Policy Act (NEPA). In practice, the individual NEPA studies are done prior to project construction, long after the SEA evaluation has been used to decide that a given project should be funded.

The specific SEA method is to quantify project costs and environmental benefits, and to derive a numeric cost-effectiveness index for each project. Projects are then ranked in order of cost-effectiveness, and the funding is allocated in accordance with the rankings until the available budget is exhausted. Some adjustments in ranking may occur for reasons other than costs and benefits, as when a particular higher-ranked project is deferred because of implementation issues, or a particular lower-ranked project is selected because a particular area of wetlands needs urgent attention. Such adjustments have been relatively minor, and the basic SEA rankings have been the primary consideration used by decision-makers to distribute about $250 million in the last eight years.

In order for this approach to SEA to work, it is necessary to have a method that provides systematic, reliable quantification of the environmental benefits that will be achieved by each project. To this end, the government agencies responsible for the funding decisions have developed a method known as the Wetland Valuation Assessment (WVA). The WVA methodology was developed by a team of wetlands scientists representing the agencies that are responsible for the funding decisions, along with experts from the academic and research communities. WVA applies the principles of habitat assessment methodology, as exemplified by the Habitat Evaluation Procedure (HEP), but with a much broader focus that reflects benefits to a variety of fish and wildlife species. This emphasis on habitat means that the method may not always capture other functional values of wetlands, such as hydrology, water quality, nutrient export, flood water storage, or storm surge protection.

In this paper, we explain the WVA methodology as an example of a tool for making strategic environmental decisions.

STRATEGIC ASSESSMENT PROCEDURES

ASSUMPTIONS UNDERLYING THE METHOD

There are two assumptions implicit in applying the WVA method for evaluation of strategic wetland benefits. First is the assumption that quantity of habitat (marsh acreage) alone is not sufficient to capture value. This reflects a general understanding that an acre of degrading marsh would not be equivalent to an acre of healthy marsh; and an acre of newly created marsh may not be equivalent to an acre of mature marsh. The method, therefore, attempts to capture the qualities that differentiate marshes in various conditions with regard to apparent functional value, specifically targeting characteristics thought to contribute to support of fish and wildlife. The method does this by using a series of variables (described below) so that the method combines estimates of quality as well as quantity.

The second assumption is that there is an optimal marsh habitat condition and it can be characterized. Wetlands quality is compared to this assumed optimal condition, and ranked accordingly.

APPROACH

There are separate WVA equations for each major wetlands type encountered along the Louisiana coast: 1) fresh/intermediate marsh; 2) brackish marsh; 3) saline marsh; and 4) cypress swamp. Each equation includes a formula to characterize the habitat quality of emergent marsh, and a formula to characterize the habitat quality of the open water areas associated with that marsh. These equations are a composite of variables that are considered important in characterizing fish and wildlife habitat. The formulas result in a number scaled from 0.1 to 1 that represents the relative value of the habitat area, compared to optimal habitat that would receive a ranking of 1. Projects are compared and funding decisions are made based on the estimate of habitat value that would be gained for each project.

Methodology — Brackish Marsh Example

Equation Variables. The approach used for assessing changes in each variable and applying these in the equation to estimate benefits will be illustrated using the brackish marsh equation. Six environmental variables are used to characterize both marsh and water conditions in the brackish (as well as the other) marsh equation(s). The main criteria used to choose equation variables are as follows:
- importance in characterizing fish and wildlife habitat quality for each wetland type
- ease in assigning values to the variable using existing information
- sensitivity to changes expected from typical wetland restoration projects
- relevance to the main coastal restoration objective of creating and preserving vegetated wetlands

The six variables are:
- percent of area covered by emergent marsh (V1)
- percent of open water area dominated by submerged aquatic vegetation (SAV) (V2)
- marsh edge and interspersion (V3)

- percent of open water that is shallow (<1.5 feet) (V4)
- salinity (V5)
- aquatic organism access (V6)

If this approach were to be applied for another purpose, for instance in a substantially different SEA or Environmental Impact Assessment (EIA) context, both the equations and the specific variables included would have to be reviewed according to the objectives of the new program or evaluation.

Suitability Indices. Each of the six variables is evaluated with regard to a Suitability Index graph. These graphs show the relationships between values of the variables and habitat quality, and are presented for the brackish marsh equation in Figure 16.1. It can be seen from Figure 16.1 that the Suitability Indices range between 0.1 and 1, where 1 represents optimal habitat. For example, the graph for V1 shows a linear relationship, where 100% marsh gives an optimal habitat suitability of 1, and no marsh (0%) gives the minimal suitability index of 0.1.

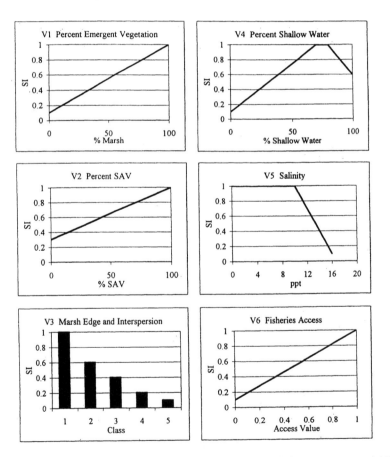

FIGURE 16.1 Suitability index graphs showing relationship of six component variables to habitat quality for the WVA brackish marsh model.

If a current value for a variable is known, the current Suitability Index can be obtained from the graph. If a future value for a variable is predicted (e.g., with or without the project), the associated Suitability Index can be obtained from the graphic relationship. The differences in these Suitability Indices then represent the changes in habitat conditions that are expected from the project, which are compared to what would occur without the restoration project.

Values for the variables included in the WVA equations are estimated through professional judgement, based on a variety of sources of existing information. These could include: recent satellite imagery (e.g., from which estimates could be made of the percentages of marsh and open water in a project area); aerial photography; field surveys conducted by the WVA workgroup; literature data; existing water quality monitoring stations and other monitoring results; interviews with experts in the field; and experience of the scientists participating in the WVA evaluation process.

The form of each Suitability Index graph reflects assumptions that were made regarding the optimal range of the variable for that marsh type, and how habitat quality would change as the value of that variable changed. As shown in Figure 16.1, the graphs for V1, V2, and V6 are a simple linear relationship to suitability — the more vegetated marsh (or SAV or organism access) there is, the higher the assumed habitat quality.

Another variable, V3 (marsh edge and interspersion), is related to suitability in a step function. Even though there is certainly a continuous range of degrees of interspersion, it was not feasible to evaluate degree of interspersion on a continuous scale.

Variable V4 (percent shallow water) is more complex. An optimal range of suitability is reached at a value less than 100% of the total possible range in values, and then declines as the value of that variable increases beyond that range. For V4, having 100% shallow water (i.e., no water deeper than 1.5 ft) was considered to reflect a lower habitat diversity and to present fewer opportunities for refuges than a combination of shallow plus deeper water.

Habitat Suitability Indices. Using the variable values estimated by the WVA workgroup, a suitability index value is obtained for each variable for existing conditions (Year 0), for the first year of project implementation (Year 1), for the end of the 20-year project life (Year 20), and for any "target year" between Years 1 and 20 for which a major change in conditions would be expected. These values are then entered into set formulas to calculate overall Habitat Suitability Indices (HSIs) for each target year. Table 16.1 gives an example analysis of a marsh creation project to illustrate the process of estimating suitability indices from variable values, combining Suitability Indices (SIs) to calculate HSIs, and then using HSIs, plus habitat acreages, to calculate Habitat Units (HUs) and then Average Annual Habitat Units (AAHUs).

TABLE 16.1
Illustration of Output of WVA Brackish Marsh Model

Project: Marsh Creation - Project X   **Project Area: 600 ac**
Condition: Future Without Project

Variable		TY 0 Value	SI	TY 1 Value	SI	TY 20 Value	SI
V1	% Emergent	10	0.19	10	0.19	9	0.18
V2	% Aquatic	5	0.15	5	0.15	5	0.15
V3	Interspersion	%		%		%	
	Class 1		0.40		0.40		0.40
	Class 2						
	Class 3	100		100		100	
	Class 4						
	Class 5						
V4	%OW <= 1.5 ft	50	0.74	50	0.74	50	0.74
V5	Salinity (ppt)	8	1.00	8	1.00	8	1.00
V6	Access Value	1.00	1.00	1.00	1.00	1.00	1.00
	Emergent Marsh HSI =		**0.37**	**EM HSI =**	**0.37**	**EM HSI =**	**0.36**
	Open Water HSI =		**0.40**	**OW HSI =**	**0.40**	**OW HSI =**	**0.40**

Condition: Future With Project

Variable		TY 0 Value	SI	TY 1 Value	SI	TY 20 Value	SI
V1	% Emergent	10	0.19	93	0.94	86	0.87
V2	% Aquatic	5	0.15	10	0.19	10	0.19
V3	Interspersion	%		%		%	
	Class 1		0.40	50	0.80	50	0.80
	Class 2			50		50	
	Class 3	100					
	Class 4						
	Class 5						
V4	%OW <= 1.5 ft	50	0.74	80	1.00	80	1.00
V5	Salinity (ppt)	8	1.00	8	1.00	8	1.00
V6	Access Value	1.00	1.00	1.00	1.00	1.00	1.00
	Emergent Marsh HSI =		**0.37**	**EM HSI =**	**0.94**	**EM HSI =**	**0.90**
	Open Water HSI =		**0.40**	**OW HSI =**	**0.49**	**OW HSI =**	**0.49**

TABLE 16.1 (CONTINUED)
Illustration of Output of WVA Brackish Marsh Model

AAHU CALCULATION - EMERGENT MARSH
Project: Marsh Creation - Project X

Future Without Project

TY	Marsh Acres	x HSI	Total HUs	Cummulative HUs
0	60	0.37	22.34	
1	59.8	0.37	22.27	22.30
20	56	0.36	20.41	405.30
			AAHUs =	**21.38**

Future With Project

TY	Marsh Acres	x HSI	Total HUs	Cummulative HUs
0	60	0.37	22.34	
1	560	0.94	526.29	227.03
20	517	0.90	465.94	9420.93
			AAHUs	**482.40**

NET CHANGE IN AAHUs DUE TO PROJECT
A. Future With Project Emergent Marsh AAHUs =   482.40
B. Future Without Project Emergent Marsh AAHUs =   21.38
Net Change (FWP - FWOP) =   **461.02**

AAHU CALCULATION - OPEN WATER
Project: Marsh Creation - Project X

Future Without Project

TY	Marsh Acres	x HSI	Total HUs	Cummulative HUs
0	540	0.40	217.57	
1	540.2	0.40	217.65	217.65
20	544	0.40	219.18	4149.85
			AAHUs =	**218.37**

Future With Project

TY	Marsh Acres	x HSI	Total HUs	Cummulative HUs
0	540	0.40	217.57	
1	40	0.49	19.78	126.31
20	83	0.49	41.05	577.89
			AAHUs	**35.21**

TABLE 16.1 (CONTINUED)
Illustration of Output of WVA Brackish Marsh Model

Future With Project

TY	Marsh Acres	x HSI	Total HUs	Cummulative HUs
NET CHANGE IN AAHUs DUE TO PROJECT				
A. Future With Project Open Water AAHUs =				35.21
B. Future Without Project Open Water AAHUs =				218.37
Net Change (FWP - FWOP) =				**–183.16**
NET BENEFITS IN AAHUs DUE TO PROJECT				
A. Emergent Marsh Habitat Net AAHUs =				461.02
B. Open Water Habitat Net AAHUs =				–183.16
Net Benefits = (2 6xEMAAHUs+OWAAHUs)/3.6				**282.08**

There are two HSI formulas for each marsh type — one characterizing emergent marsh habitat quality, and one characterizing open water habitat quality. The formulas for brackish marsh are as follows.

$$Emergent \text{ Marsh HSI } = \frac{\left[3.5*\left(SI_{V1}^5 * SI_{V6}^{1.5}\right)^{1/6.5}\right]+\left[\left(SI_{V3}+SI_{V5}\right)/2\right]}{4.5}$$

$$Open \text{ Water HSI } = \frac{\left[3.5*\left(SI_{V2}^3 * SI_{V6}^2\right)^{1/5}\right]+\left[\left(SI_{V3}+SI_{V4}+SI_{V5}\right)/3\right]}{4.5}$$

In each formula, the variables are weighted using coefficients, or exponents, or both. Selection of the coefficients and exponents reflects judgments on the relative importance of each variable in determining habitat quality for that wetlands type. Overall, the coverage of marsh (V1) is considered the most important variable. Thus, V1 (emergent marsh) and/or the term containing V1 typically has the largest weighting factor. The absolute values of the weighting factors were not based on specific ecological principles, but were selected to achieve a comparative status that reflects a consensual judgement of importance made by the WVA workgroup members. For example, in the first term of the brackish marsh formula, V1 (emergent marsh) was weighted to control at least 50% of the variability in the marsh HSI formula. The actual value used (e.g., an exponent of 5) to achieve the defined relative level of importance was in large part influenced by the number of variables included in the equation. V2 (submerged aquatic vegetation) and V6 (organism access) were judged to carry approximately equal weight in the brackish open water formula, and in turn, were considered more important than V3 (interspersion), V4 (shallow water), or V5 (salinity). In each case, the denominators were defined by the combination of weighting factors applied in the numerator, so that the final quantity still ranges between 0.1 and 1.0. The HSIs are assumed to have a linear relationship to overall ability of the particular marsh or water area to provide fish and wildlife habitat.

Habitat Units. As illustrated in Table 16.1, the next step in estimating project benefits is to multiply each marsh HSI by the acres of marsh present in each target year, and to multiply each open water HSI by the acres of open water in each target year. This yields HUs of marsh and of open water for each target year. HUs can be thought of as acres modified by a quality factor.

Marsh HUs are then averaged over the 20-year project life; the same is done for open water HUs. This gives AAHUs for both marsh and water for the future with the project and for the future without the project. AAHUs for future without the project are subtracted from AAHUs for future with the project to estimate net AAHUs for marsh and for water habitat due to the project.

To get total project benefits, AAHUs for marsh and water habitats are combined using formulas that, again, emphasize the relatively greater importance attributed to emergent marsh. The formula for brackish marsh is as follows.

Brackish Marsh:

Total Benefits = [(2.6 * Emergent Marsh AAHUs) + Open Water AAHUs]/3.6

Average annual habitat units are combined with annualized project cost to define cost-effectiveness, and from this a relative ranking for each project. Rankings are modified slightly based on factors that reflect consistency with long-term cost effectiveness and restoration program goals such as longevity, sustainability, and implementability.

Methodology — Other Marsh Types

The overall process illustrated in the brackish marsh example (above) is the same for fresh/intermediate and for saline marshes. The component variables are also the same. However, the SI graphs that relate values of each variable to habitat quality differ between marsh types, reflecting expected differences in the optimal range for some variables, especially salinity and abundance of submerged aquatic vegetation, between marsh types.

In addition, the formulas used to calculate HSIs for marsh and open water in fresh/intermediate and saline marsh differ slightly from those for brackish marsh. The main differences are in the relative weightings of each variable. These differences reflect judgements on how important each variable is in determining habitat quality for that marsh type. For instance, abundant SAVs (V2) are considered more a characteristic of fresh marshes than saline marshes. Conversely, the fresher a marsh is, the less influential estuarine organism access is considered to be. The formulas for fresh/intermediate and saline marshes are as follows.

Fresh/Intermediate Marsh:

$$\text{Emergent Marsh HSI} = \frac{\left[3.5*\left(SI^5_{V1}*SI^1_{V6}\right)^{1/6}\right]+\left[\left(SI_{V3}+SI_{V5}\right)/2\right]}{4.5}$$

$$\text{Open Water HSI} = \frac{\left[3.5*\left(SI^3_{V2}*SI^1_{V6}\right)^{1/4}\right]+\left[\left(SI_{V3}+SI_{V4}+SI_{V5}\right)/3\right]}{4.5}$$

Saline Marsh:

$$\text{Emergent Marsh HSI} = \frac{\left[3.5*\left(SI^3_{V1}*SI^1_{V6}\right)^{1/4}\right]+\left[\left(SI_{V3}+SI_{V5}\right)/2\right]}{4.5}$$

$$\text{Open Water HSI} = \frac{\left[3.5*\left(SI^1_{V2}*SI^{2.5}_{V6}\right)^{1/3.5}\right]+\left[\left(SI_{V3}+SI_{V4}+SI_{V5}\right)/3\right]}{4.5}$$

Finally, the formulas used to combine marsh and open water AAHUs to obtain final net project benefits differ slightly between marsh types. The formulas for fresh/intermediate and saline marshes are as follows.

Fresh Marsh:

Total Benefits = [(2.1 * Emergent Marsh AAHUs) + Open Water AAHUs]/3.1

Saline Marsh:

Total Benefits = [(3.5 * Emergent Marsh AAHUs) + Open Water AAHUs]/4.5

APPLICATION AND EFFECTIVENESS OF THE METHOD

The WVA is a method of quantifying a series of subjective judgements on wetlands value and condition, so that benefits of alternative projects can be quantified and compared. As experience of WVA workgroup participants has increased over eight years of applying the WVA methodology, the application of judgment has become more consistent, with more efficient use of both existing information and field surveys of all proposed project sites. The results of these model applications thus have become increasingly well regarded.

At the time of writing of this chapter, the WVA methodology had been applied to the selection of 55 projects on 6 of 7 annual project lists. (The method had not been fully developed at the time of selection of the first annual project list; and an eighth list of candidate projects has been evaluated and ranked, but not yet selected for funding.)

The effectiveness of the method is not easy to quantify. Ranking based on WVA benefits, plus cost, has played the major role in evaluating and selecting restoration projects valued at more than $250 million. But other considerations have played a role in final project selection. The additional factors considered in establishing final project selection have changed over the years as the restoration planning process has matured and provided greater guidance to initial identification of candidate projects for review, and as greater efforts have been applied to initial project development. In general, other considerations fall into three categories: suitability for long-term restoration, practical issues regarding implementability, and political considerations.

Additional, or secondary, factors that were considered in the final selection of projects for the second through the fourth annual project list include: degree of support from the State of Louisiana, local government and landowners, and the public; the degree to which the project introduces fresh water and sediments; rarity of the project's marsh type within the basin; association of the project with other environmental programs; project benefits to threatened or endangered species; criticality of the project to overall basin restoration; and implementation issues, such as anticipated conflicts with land rights or oyster lease issues. No quantitative formula was used to incorporate these factors into the ranking and project selection process. Rather, subjective judgments on these factors was considered in addition to the WVA benefit/cost ranking, and participating federal and state agencies voted on project funding priorities.

In subsequent years (i.e., for the 1995 to 1997 annual project lists), a modified set of secondary criteria were incorporated into a formula that also included the WVA benefit/cost results, to give a final ranking to candidate projects. The modified secondary criteria included: degree to which the project provides for long-term, sustainable restoration; degree to which the project supports the restoration plan strategies developed for that region; degree to which the project is supported by non-federal funding partnerships; degree of public support; and the degree of risk or uncertainty associated with the proposed project. While judgements on each of these criteria remain subjective, the weight given to each criterion compared to that applied to the cost-effectiveness ranking is set by the formula. Thus, application is more consistent among projects.

In both cases, these secondary considerations could cause a project ranked higher based on WVA benefits/cost to be passed over for funding, allowing a lower ranked project to be considered. Or, a lower ranked project could be targeted for funding based on secondary considerations, displacing other more cost effective projects. Based on our review of actual rankings, where a project was highly ranked based on the WVA, the project was almost always funded; the principle exceptions were projects that faced practical obstacles to implementation. Similarly, projects that were ranked very low on the WVA were seldom funded; the principle exceptions were where the funding agencies perceived that a given habitat had special value not reflected in the ranking (e.g., barrier island restoration).

The most common departure between funding decisions and WVA rankings occurred for projects of intermediate ranks, which were near the edge of the funding cut-off. The last few projects added to any given priority list were often not the best remaining projects from a WVA perspective. In part, this reflected a judgement that the WVA results are not precise, so that small to moderate differences in WVA rankings were not considered as definitively showing that one project has more benefits than another.

ACKNOWLEDGMENTS

Many scientists participating in the CWPPRA contributed to development of the WVA models. The U.S. Fish and Wildlife Service (USFWS) took the lead in model development. Lloyd Mitchell of USFWS was the prime mover in initial model development, while Kevin Roy of USFWS and current chair of the WVA workgroup played the primary role in subsequent model revisions. Dr. Gary Shaffer of the University of South Louisiana conducted a significant statistical evaluation of the WVA model forms and component variables. Active members of the WVA workgroup, who have all participated in model evaluation and revision, include Jeanene Peckham of the U.S. Environmental Protection Agency; Sue Hawes of the U.S. Army Corps of Engineers; Marty Floyd of the U.S. Natural Resource Conservation Service; Rick Hartman of the National Marine Fisheries Service; and Darryl Clark of the Louisiana Department of Natural Resources.

17 SEA: The Principles of Negotiation, the Disaggregative Decision-Making Method, and Parallel Organization in Regional Development

Heli Törttö

CONTENTS

INTRODUCTION

The Brundtland Commission drew the world's attention to environmental aspects of political decision-making in 1987.[1] In 1992, the United Nations Conference on

Environment and Development agreed that environmental impacts of both projects and policies, plans, and programs (PPP) should be assessed to improve the quality of decision-making.[2]

Over the last decade, demands for integrating environmental aspects into the policy of the European Union (EU) have increased. The Single European Act (1987) supports this development, as environmental protection became one of the objectives of the EU.[3] In the articles of the Maastricht Treaty (1992), which are specific to environmental issues, sustainable economic development was agreed to be one of the objectives of the EU.[4] The main theme of the Fifth Environmental Action Program (1993-1997) is sustainable development.[5]

There is an international agreement that environmental aspects should be taken into consideration in political decision-making, but development of procedures has been limited. Meeting the set requirements varies greatly, for example, between the different Member States of the EU:

> Notwithstanding the fact that the primary responsibility for implementing environmental and cohesion policy rests with the Member States, the Commission has for several years been receiving complaints concerning infringements of environmental legislation in the implementation of projects assisted by Community funds. The Commission views this situation seriously in that it damages public perception of Community Activity.[6]
>
> During the first round of Structural Funds programming (1989–1993) the European Parliament, but also the Court of Auditors and environmental non-governmental organisations expressed criticism on the lack of systematic environmental appraisal procedures in the programming, as well as on cofinancing of projects allegedly damaging the environment.[6]
>
> The continuing challenge is to ensure that the implementation of these programs is consistent with sustainable development and Community environmental rules.[6]

Integration of environmental concerns into political decision-making is seen as a sensible way to the implementation of sustainable development, creating opportunity for welfare in the long run. Resources to solve problems after they have occurred are not needed. It is often easier to prevent problems from occurring in advance than to deal with them once they have happened.

Sustainable development has become a slogan of today's society. To achieve sustainable development, wide knowledge, as well as open information and discussion when preparing decisions, are needed. Environmental impact assessment (EIA) is seen to facilitate those conditions, and as such, it is one way to achieve sustainable development.

The purpose of this chapter is to demonstrate that Strategic Environmental Assessment (SEA) can play a significant role in the implementation of sustainable development, and that, to that end, the elements of SEA must include, among others, cooperation between different sectors of society, public participation, use of disaggregative decision-making methods, and comparison between alternative future trends.

FRAMEWORK OF THE STUDY

In the research completed at the University of Oulu, the aim was to find ways to integrate SEA into political decision-making at the regional development level. In the Finnish case that will be referred to in the chapter, the Northern Ostrobothnia Regional Development Programs means the production of regional development programs at the national level and the EU Structural Funds Objective 5b and 6 programs. Projects are assisted by regional development appropriation at the national level, including regional development funds and financing from the EU Structural Funds, as shown in Figure 17.1. Financial aid through these programs is granted for developing, for example, productive business, tourism, and the countryside.

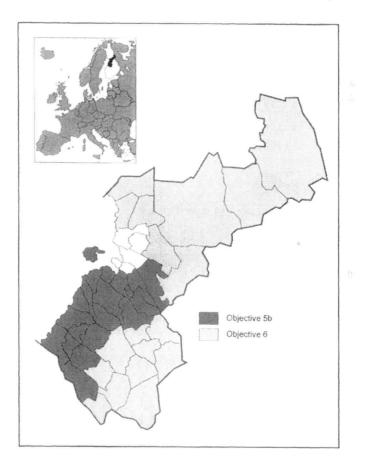

FIGURE 17.1   Eligible regions in Northern Ostrobothnia according to the EU Structural Funds 1994–99 (from the Regional Council of Northern Ostrobothnia. With permission).

The question was: what kind of planning is required for an effective application of the above mentioned principles of SEA? Leskinen (1994) has made a distinction between routine planning, which represents the natural science approach, and planning as learning, representing the human science approach, as shown in Figure 17.2. In that theoretical framework, the prevailing natural science approach in social sciences is questioned, and a human science approach is proposed as a basis for environmental planning. The key criteria of environmental planning becomes the ideal of participative democracy, the ability to learn from experience. As such, Leskinen's planning as a learning option can be seen to faciliate the adoption of the principles of SEA.

Leskinen's distinction between routine planning, representing the natural science approach, and planning as learning, representing the human science approach, were considered as analytical tools when the regional development process in Northern Ostrobothnia was studied. The ultimate aim was to solve the main environmental problem in Northern Ostrobothnia that culminated in cumulative impacts of small projects.

The analysis demonstrates that the main principles of public participation in a first stage of program development were those typical of routine planning: informing and two-way exchanging of information, but that the principles of planning are increasingly changing to planning as learning. On the other hand, the main principles of decision-making were those typical of routine planning, meaning, planning did not proceed through to alternatives. In fact, some of the basic principles of routine organization were detected throughout the process.

As a result, it is argued that a different form of planning is needed, which is based on planning as learning approaches. The concept of sustainable regional development is proposed.

BACKGROUND OF THE STUDY

The use of natural resources in Northern Ostrobothnia is typically the result of the implementation of small projects, as demonstrated by the following examples:

> In 1994, 30 274 hectares of forest were cut down in the area of the Forestry Board of Northern Ostrobothnia, the total number of projects was 3 500. Forest drainage was completed in the area of 12 525 hectares, the total number of projects was 175.[7]
>
> In 1992, there were 15 000 hectares of peatland under peat production, the total number of producing peatlands was 101 in the area of the Northern Ostrobothnia Regional Environment Centre. In addition to that, the total number of peatlands ready for peat production was 39.[8]

These projects have accumulated effects in water systems, resulting in environmental impacts, which are difficult, or even impossible, to detect and control on a case by case permitting system.

In addition to forestry and peat production, the rivers in Northern Ostrobothnia have high nutrient load caused by agriculture:

> In 1992, the number of farms in forestry and agriculture was 9851. Agriculture was mainly in the form of cattle farming, 62% of the farms were cattle farms. These farms had 191 200 hectares of land; an average of 19.4 hectares per farm.[9]
>
> In the rivers of Northern Ostrobothnia agriculture is the main reason for nitrogen and phosphorus load. 80% of agriculture is centred in river valleys. Surface runoff is

ROUTINE PLANNING	PLANNING AS LEARNING
INFORMING OR TWO-WAY EXCHANGE OF INFORMATION	**NEGOTIATION**
Planning is expert work. Citizens are recipients and, to some extent, producers of information.	Planning consists of systematic discussions between experts and parties concerned.
The organization decides on the content of the plan, the participants, and the different values to be included. The planning process is closed.	All parties can take part and agree on the rules of negotiation and the contents of the decision-making material. Values of all parties are shown. The planning is open.
AGGREGATIVE DECISION-MAKING METHOD	**DISAGGREGATIVE DECISION-MAKING METHOD**
Renders all impacts commensurate. The feasibility of each option is shown by combining its impacts in a numerical value.	Describes impacts separately and in different ways. Versatility and intelligibility are aimed for in order to enable the discussion of options.
The decision-making process and expert are considered to be basically value-fee, and the choices made are thus thought to be objective.	Expert and method are value-bound; openness and clear reasoning are therefore essential.
The aim is to select the best option for attaining given goals.	Various goals and options are defined. The method describes the options from the point of view of various goals. The choice between options is based on the rules of participative democracy (not by the decision-making method).
ROUTINE ORGANIZATION	**LEARNING ORGANIZATION**
Changing its statutory functions is not the organization's business.	Statutory functions need reinterpreting. Cooperation with other sectors of administration is necessary even if not statutory.
The main aims are techno-economic efficiency and the observation of laws and norms.	Efficiency is not a straightforward matter; partial, techno-economic optimization is not enough. Norms are taken to be relatively permanent, but subject to discussion.
The structure is hierarchical - instructions and orders in standard format are sent downward and control reports upward.	The organization adjusts its activities and structure to each situation. Only partly standardized information flows horizontally and vertically.
Loyalty and obedience to the rules are the main virtues expected of the planner. Development of the organization is the management's business.	Everyone can take part in developing his organization. The aim is to make use of the experience and views of all those concerned as comprehensively as possible.

FIGURE 17.2   The main options of environmental planning, "routine planning", and "planning as learning" (from Leskinen, A., *Environmental Planning as Learning: The Principles of Negotiation, the Disaggregative Decision-Making Method and Parallel Organization in Developing the Road Administration*, University of Helsinki, Helsinki, 1994. With permission).

. high because the country is low-lying and permeability of the soil is poor. In terms of the amount of nutrients, the river valleys in Northern Ostrobothnia are the most efficient areas under agriculture in the whole of Finland.[9]

The importance in controlling cumulative impacts in Northern Ostrobothnia cannot only be seen to be dependent on the structure of trade but also on the structure of human settlements. In the Northern Ostrobothnia region, settlements tend to accumulate in the Oulu region and in the centers of municipalities, while settled areas in the countryside tend to decline:

> The area that is settled in Northern Ostrobothnia has declined, though the number of inhabitants has increased. If the settled areas were to be divided into metre square boxes, more than half of them would be loosing inhabitants. When looking at the municipalities, in 1970–80 only the centres of the municipalities have grown, and after this period also the surrounding small villages.[10]

Typical to the settled areas is also the deconcentration of the settlements in the Oulu region:

> The Oulu region has been the most rapidly growing area in Finland. Since 1970, the number of inhabitants has increased by 46 000. The settled area has spread and the density has remained low. Working places are still in the centre of Oulu town. Distances between homes and the time spent travelling to and from work have increased. In 1990, almost 15 000 workers travelled to Oulu to work and the distance was on average almost 19 km.[11]

This all leads to growing expenses and increased traveling. It is not easy to take measures to prevent this development. To solve the problems above, environmental concerns should be taken into consideration right at the strategic level of political decision-making.

Traditional planning approaches have to change to enable environmental concerns to be integrated into political decision-making. SEA can be a very useful approach. In this process, the critical elements of SEA include, among others, cooperation between different sectors of society, openness of planning, and a chance of public participation. Comparisons between alternative future trends, multiple and clear picturing of impacts, use of disaggregative decision-making methods, and use of quantitative and qualitative information are needed.

In SEA, emphasis is put on the impacts that cannot be perceived at the project level. These are, for instance, cumulative impacts of small projects. It is argued that SEA should be an integral part of planning. The success of planning can then be measured by using the information obtained in monitoring and auditing programs, thus building on experience that will be used in future planning.

An organization engaged in environmental planning must decide on how to cope with the following issues:

1. Contacts with citizens and interest groups: choosing among various principles of participation;

2. Gathering and processing of information: choosing among various principles of decision-making methods;
3. Internal flow of information and contacts with other public bodies: choosing among organizational principles.[12]

These issues form the basis for the discussion of the argument as follows.

CONTACTS WITH CITIZENS AND INTEREST GROUPS, INTERNAL FLOW OF INFORMATION, AND CONTACTS WITH OTHER PUBLIC BODIES

PROGRAM PREPARATION

In the General Agreement of the Regional Council of Northern Ostrobothnia, the regional council is committed to act as: A regional cooperation organization.[13]

Likewise, in the Northern Ostrobothnia Regional Development Program, the regional council's nature of a cooperation organization is emphasized.

Despite this fact, at the first stage of program development, the principles of planning were those typical of routine planning. The first regional development program was based on the development programs prepared by the sub-regions. Public participation was broad and interactive in the sub-regions. The development ideas and projects of the sub-regions were gathered together at the regional level. At that stage, the main principles of public participation were those typical of routine planning: informing and two-way exchange of information.

There is an effort however to increasingly adopt planning as learning in Northern Ostrobothnia. Systematic cooperation between experts and other involved parties, active interaction, and openness are being emphasized. The regional development program now offers an open forum for different views and development ideas to be presented and discussed. Sustainable development, as a common issue and objective, becomes a common ground for reconciling different views and ideas, also opening opportunities to bring environmental and welfare considerations together.

In 1995, the Board of the Regional Council of Northern Ostrobothnia indicated that the regional council should have a Strategy of Action. In this strategy, emphasis is put on both internal ways of action and external interaction. The vision is that the regional council will act as the center of cooperation and development:

> The Regional Council of Northern Ostrobothnia is an eligible co-operation companion and a trusted flagship which shows the course and seizes new challenges actively, directing skilful actors of the region to work together for the better of Northern Ostrobothnia.[14]

The meaning of cooperation is emphasized in all actions of the Regional Council of Northern Ostrobothnia: The role of the regional council would be the co-ordinator, parturient and leader of co-operation.[14] The Regional Council of Northern Ostrobothnia is a co-operation organisation. Well functioning co-operation between interest groups will ensure fruitful regional development.[14]

This forms the basis for the revision process of the regional development programs. The aim of the revision process calls for as wide a participation, as possible, of all parties that implement the programs (authorities, sub-regions, companies, educational establishments, and other interest groups) and persons that have been elected to a position of trust. The role of the board of the regional council is seen to be active. New persons elected to a position of trust will participate in the process all the way through. The outcome will be: Development program produced and accepted by the region.[15]

The revision process will be supported by continuous informing of interest groups and the media. The main principle is that: The revision process is a common project of the region whose foundations are active cooperation, interaction and openness.[15]

PROGRAM IMPLEMENTATION

Implementation of the programs contains elements of routine planning. Techno-economical effectiveness is a central objective. The Regional Management Committee, which is in a central position in financial decision-making, is given a short summary of each project containing the following information: names, central contents, objectives, and outcomes of the proposed projects, statements of the reasons for the proposed projects, and draft resolutions. There is no examination on the environmental impacts of the proposed projects.

When it comes to business involvement, chances of the Regional Management Committee to take a stand on environmental impacts are small. Projects handled in the Business Aid Working Group come as a notice to the Regional Management Committee. These notices include the names of the enterprises receiving funding and the amount of funding appropriated. Any other information is considered to be commercially sensitive.

Regional development authorities are represented in the Regional Management Committee Working Group and in the Business Aid Working Group. Environmental authorities are not represented in either of the Groups, which is problematical from the environmental point of view, especially as important decisions concerning environmental issues are made at this stage.

The majority of the funding decisions concerning regional development projects are made outside of the Regional Management Committee. The Regional Management Committee makes decisions on the regional development funds and financing from the EU Structural Funds. During 1994 and 1995, the regional development funds in Northern Ostrobothnia amounted to 1.7 million ECU. Financing from the EU Structural Funds were 3.8 million ECU. The total amount of regional development appropriation was 266.7 million ECU. Projects assisted by the regional development appropriation are not justified by the measures and the priorities presented in the national regional development program. Each sectoral authority justifies funding without any coordination.

PROGRAM MONITORING

Monitoring contains elements of routine planning. Monitoring is coordinated by the Regional Management Committee. The authorities who have been nominated to act as supervisors respond for the monitoring of the funded projects. Supervisors take part in the meetings of the monitoring committees and they follow up the costs of the projects. The final aim is techno-economic effectiveness.

GATHERING AND PROCESSING OF INFORMATION

PROGRAM PREPARATION

At the first stage of program development, principles of planning were those typical of routine planning. Different alternatives were not examined. The measures presented in the programs are broad and they can be considered to contain the development possibilities of Northern Ostrobothnia.

Environmental impacts were presented, but verbally, measure by measure, and their costs were not internalized. Quantitative EIA would be difficult in any case. There is a lack of information concerning, for instance, impacts caused by agriculture and forestry on waters. EIA on a project by project basis is not a good solution, because only a fraction of the projects presented in the regional development programs were funded during 1994 and 1995.

On the other hand, elements of planning as learning can also be detected. The first phase was a strategic education program organized by the University of Oulu. During this program, different sub-regions, future prospects were examined by using scenario analysis.

In the revision process of the regional development program, sustainable development is the main axe of the program. For example, when searching for measures to gain direct and indirect positive impacts on employment, the criteria is that those projects lay a sustainable basis for the success of the region in the long run. In other words, the projects preserve the vitality of the environment (natural resources, environment load, biodiversity) and a well-balanced progress of welfare (standard of living, quality of life, cultural heritage). At the first stage, different future development alternatives are created. They are analyzed by using the criteria for sustainable development. Strategic environmental objectives are chosen and measures are set on the basis of the alternatives and their assessment. The aim is that the revised program would be: "The program of sustainable development."[15]

PROGRAM IMPLEMENTATION

During the years 1994 and 1995, regional development funds have mainly been used for financing research projects and projects improving the preconditions of the operation of economic life. These projects are said to not have significant environmental impacts.

However, projects that are handled by the Business Aid Working Group are likely to have environmental impacts. Environmental impacts of these projects are hard to assess because information concerning these projects is considered to be commercially sensitive. The amount of regional development appropriation directed to business aid was notable: 69% (41.4 million ECU) in 1994 and 44% (80 million ECU) in 1995. This support was applied to different lines of business, like food production, wood processing, activities supporting tourism, paper products, products of metal, and machine workshops and electrical products.

In 1995, 1.3% (2.3 million ECU) of the regional development appropriation was directed to water supply, environmental protection of islands and mountains, and water protection; 3.2% (5.8 million ECU) was directed to forestry, i.e., forest cultivation, renovation ditching, and road building. Maintenance of public roads, development of road nets, and building and maintenance of aviation areas received 41% (74.8 million ECU). Building channels, minor harbors and maintenance, and development of railways was granted 1.3% (2.3 million ECU). Finally, 1.7% (3.2 million ECU) was granted for public transport.

There are no specific project selection criteria for the projects assisted by regional development funds. Regional development funds can be granted to carry out the development objectives of the region. In Northern Ostrobothnia, these objectives include unemployment reduction by creating new jobs and by protecting the existing ones. A central principle for achieving such objectives is to promote sustainable development.

Project selection criteria for projects assisted by the EU Structural Funds are usually related to economic activity. When funding is given, it is traditionally discussed from the point of view of economy, employment, trades, and competitive ability. The criteria are, for example: "Improvement of employment in the area."[16] "Improvement in productivity."[16] "Improvement of the attractiveness of the region for business."[16]

The environmental project selection criteria are general, like: "Improvement in environmental protection and environmental technology."[16]

During the years 1994 and 1995, environmental impacts of projects assisted by the regional development appropriation and the EU Structural Funds included: 1. Primary production: water load caused by agriculture, forestry, and peat production; 2. Refinement: threats to biodiversity and environmental load caused by energy production; 3. Consumption: environmental load caused by growth of traffic, tourism, and recreation. No project selection criteria concerning these matters were presented in the regional development programs.

Funding decisions are made one at a time. Cumulative impacts are not taken into consideration. In 1994 and 1995, 12 projects concerning peat production were assisted by the regional development appropriation in Northern Ostrobothnia. Nine of these projects were situated in the river Iijoki watershed. In 1995, 313 projects concerning land drainage for foresting were assisted by the regional development appropriation in Northern Ostrobothnia. Of these projects, 67 were situated in the river Iijoki watershed.

PROGRAM MONITORING

There can also be detected elements of routine planning in the monitoring of the regional development programs. The main objective of monitoring is techno-economic efficiency. Projects assisted by the regional development funds are monitored by following the costs of the projects. Environmental impacts are not monitored.

Indicators presented in the EU Structural Funds programs mainly represent aspects of economic activity. Indicators are quantitative, and they are related to, for example, reducing unemployment and creating jobs, raising the Gross National Product (GNP), the number of business companies established, and turnover of industry: "Overall increase in tourism."[16] "Number of small and medium-sized enterprises created."[16]

Indicators relating to environmental issues are very general and include: "Reduction of adverse environmental effects."[16] "Fisheries improvement in estuaries and rivers."[16]

Regional development is a continuous process, whereby the results of previous experiences should be taken into consideration in future decisions. Through monitoring it should be ensured that the projects funded through regional development programs, and their impacts, do not conflict with the strategies presented in future programs. On the other hand, the information gained with monitoring could also be used in the future rounds of planning and beyond. In other words, organizations should be able to learn.

DEVELOPMENT RECOMMENDATIONS

Sustainable regional development is thought of as a form of planning that integrates SEA into regional development. In this model, both operational sectors of the regional councils, regional development and regional planning, fulfill the desired prospect for the future defined in the regional strategy. Sustainable development is taken into account in the regional strategy and in the preparation, implementation, and monitoring of the regional development program.

Regional councils are the centers of development and cooperation. The main objective is to get the regions to work as active and successful sets. Regional development programs offer a forum for this purpose. On the other hand, sustainable development acts as a benchmark for regional development. The development visions of the regions, their strategies and measures, are chosen in a manner that establishes preconditions for sustainable development. In other words, the vitality of the environment and the well-balanced progress and welfare are ensured.

The central elements of sustainable regional development are the following:

To assist contacts with citizens and interest groups, internal flow of information and contacts with other public bodies:

1. *sustainable development strategy of action* based on the information on the state of the environment
2. *regional strategy* including the identification of development alternatives of the region

3. *project selection criteria and monitoring indicators* based on the sustainable development strategy of action
4. *sustainable development monitoring program and sustainable development monitoring report*
5. *environmental impact assessment program*

To assist gathering and processing of information:

6. *sustainable development expert group* would help to prevent the problems rising from the sectoral environmental administration characteristic in Finland. The task of this expert group would be to perceive and solve problems related to sustainable development together with administration in program preparation, implementation, and monitoring
7. *negotiating committees of sub-regions to promote public participation*
8. *participation programs* both at the regional level and at the sub-regional level to prevent problems rising from not planning participation beforehand
9. *training programs* for regional councils and public organizations concerning new tasks and cooperation in connection with sustainable regional development.

CONCLUSIONS

Systematic cooperation between experts and other parties involved in planning, open planning, and wide participation were emphasized in SEA. Examination of different alternatives and objectives, multiple and clear descriptions of impacts, and open comparisons of alternatives are also emphasized. Impacts that cannot be considered only at the project level exist while EIAs are considered to be important. Key impacts result from cumulative impacts of small individual projects. Information on the degree of success, and need for SEA, results from monitoring and auditing, which is used to check on past experiences and used for future assessments. SEA is considered to be an important part in preparing decisions.

In Northern Ostrobothnia, solving environmental problems is highly related to taking cumulative impacts into consideration. Regional development of Northern Ostrobothnia was chosen to be the case study of the research recently carried out at the University of Oulu. Principles of routine planning and planning as learning were used as analytical tools when SEA in regional development was analyzed. The principles of planning in Northern Ostrobothnia are changing more and more to planning as learning. The role of the Regional Council of Northern Ostrobothnia as a center of development and cooperation is based on the fact that the regional council is responsible for both regional and land-use planning and, since 1994, also regional development. Regional development programs offer a forum for different opinions and development ideas. Sustainable development is the basis to make them compatible. Sustainable development can be considered to be a bridge that helps to understand and combine environmental and welfare considerations.

At the first stage of program development, principles of planning were those typical of routine planning. The first regional development programs were based on the regional development programs made in the sub-regions. In the sub-regions, public participation was broad and interactive. The development ideas and projects of the sub-regions were gathered together at the regional level. At that stage, the main principles of public participation were those typical of routine planning: informing and two-way exchange of information. Different alternatives were not examined. The measures presented in the programs are broad and they can be considered to contain the development possibilities of Northern Ostrobothnia.

In the revision process of the national regional development program sustainable development is the main axes of the program. When searching for measures to gain direct and indirect positive impacts on employment, the criteria are that projects create sustainable substrates for the success of Northern Ostrobothnia in the long run. In other words, they preserve the vitality of the environment and well-balanced progress of welfare. At the first stage, different views of future development possibilities are created. Development alternatives are analyzed by using criteria for sustainable development. Strategic environmental objectives are chosen and measures are set on the basis of alternatives and their assessment. The aim is that the revised program is the program of sustainable development.

The revision process of the national regional development program and the Strategy of Action of the Regional Council of Northern Ostrobothnia both emphasize the principles of planning as learning. In other words, there's cooperation between different sectors of society, a chance for public participation, and openness. However, special attention should be paid to the environmental impacts of the projects assisted by the regional development appropriation. Some of the basic principles of routine organization can be detected. To be able to follow the principles of planning as learning, organizational principles should be changed, especially in the implementation and monitoring of the regional development programs. Typical to the organizational principles is techno-economical effectiveness.

Sustainable regional development is proposed to integrate SEA into regional development. In this model, both operational sectors of regional councils, i.e., regional development and regional planning, fulfill the desired future prospect defined in the regional strategy. Sustainable development is taken into account in the regional strategy and in the preparation, implementation, and monitoring of the regional development program.

REFERENCES

1. World Commission on Environment and Development, *Our Common Future,* Oxford University Press, London, 1987.
2. The United Nations Conference on Environment and Development (UNCED), *The Earth Summit,* Graham & Trotman Ltd., London, 1993.
3. European Communities Commission, *Single European Act,* Bulletin of the European Communities, Supplement 2/86, Office for the Official Publications of the European Communities, Luxembourg, 1986.

4. Council of the European Union/Council of the European Communities, *Treaty on European Union,* CB-NF-86-002-EN-C, Office for Official Publications of the European Communities, Luxembourg, 1992.
5. Directorate General for Environment, Nuclear Safety and Civil Protection, *Towards Sustainability, an European Community Program of Policy and Action in Relation to the Environment and Sustainable Development,* CR-77-92-142-EN-C, Office for Official Publications of the European Communities, Luxembourg, Brussel, 1993.
6. Commission of the European Communities, *Cohesion Policy and the Environment,* Communication from the Commission to the Council, the European Parliament, the Economic and Social Committee, and the Committee of the Regions, COM(95) 509 final, 1-2, 5, 1995.
7. Forestry Board of Northern Ostrobothnia, *Annual Report 1994,* Oulu, 6, 9, Table 4, 1994. Finnish original, translation by the author.
8. Northern Ostrobothnia Regional Environment Centre, unpublished data, 1995. Finnish original, translation by the author.
9. Regional Council of Northern Ostrobothnia, *Northern Ostrobothnia Regional Development Program 1995-1999, Countryside Program,* Publications of the Regional Council of Nothern Ostrobothnia A:3, Oulu, 8, 10-11, 1994. Finnish original, translation by the author.
10. Regional Council of Northern Ostrobothnia, *The Vitality of the Villages in Northern Ostrobothnia,* Publications of the Regional Council of Northern Ostrobothnia A:13, Oulu, 12, 1995. Finnish original, translation by the author.
11. Regional Council of Northern Ostrobothnia, *Analysis of the Community Structure in the Oulu Region,* Publications of the Regional Council of Northern Ostrobothnia A:11, Oulu, 6, 10, 14, 32, 34, 1995. Finnish original, translation by the author.
12. Leskinen, A., *Environmental Planning as Learning: The Principles of Negotiation, the Disaggregative Decision-Making Method and Parallel Organization in Developing the Road Administration,* University of Helsinki, Department of Economics and Management, Land Use Economics, Publications No. 5, Land Use Economics, Helsinki, 31, 1994.
13. Regional Council of Northern Ostrobothnia, *General Aggreement of the Regional Council of Northern Ostrobothnia,* 1. Finnish original, translation by the author.
14. Regional Council of Northern Ostrobothnia, *Northern Ostrobothnia Strategy of Action,* Publications of the Regional Council of Northern Ostrobothnia, B:10, Oulu, 6, 8, 10, 1996. Finnish original, translation by the author.
15. Turunen, T., Personal communication, 1997.
16. European Commission, *Finland; Single Programming Document 1995-1999; Objective 6,* No Feder 951413002, No Arinco 95F116002, 72, 74-75, 78, 81, 103, 139, 1995.

18 SEA: The Context for Ex-ante Public Participation in Transportation Planning: The Quebec Experience (Canada)

Pierre André and Jean-Pierre Gagné

CONTENTS

INTRODUCTION

During the 1950s and 1960s, the metropolitan areas of Quebec underwent a substantial urban sprawl, which was generated by and has required the construction of different highways. Both the structuring effect of highways, namely on the residential growth, and the increase of environmental concerns have contributed to the emergence of specific environmental concerns among people living near the highways. They argue that the quality of their environment has been deteriorated by increases in noise, dust, visual degradation, and air pollution; among these factors, noise occupies a prime position. They have stated that such degradation contributes to a reduction of the price of their property.

1-56670-360-3/00/$0.00+$.50
© 2000 by CRC Press LLC

The degradation of the sound environment contributes to impacts on health and human behavior. In an exhaustive literature review, André and Gagné[1] have noted the following health impacts of noise: an increase of cardiac beats and blood pressure, the degradation of moods, an increase in the use of sleeping pills, a tendency in the most sensitive people to develop psychological distress and feelings of depression, an increase of headaches and digestion problems, an increase in the secretion of different hormones and neurotransmitters, and, in the most severe cases, a deterioration of the hearing acuity. Noise also contributes to a significant degradation of the subjective and objective quality of sleep. In terms of impacts on behavior, they noted an influence on the performance at work, on the learning process at school, on the communication process, on relaxation activities, on the social relationship, on investments in technology (not only noise-control related), on the apartment layout and related activities, and on the general mobility of the residents. Many factors may modulate these impacts including age, sex, mental health condition, hearing acuity, noise sensibility, and other cultural and socio-economic conditions. Even if we cannot generalize for all the people, it is clear that, according to the literature available, the noise needs to be considered, predicted, reduced, mitigated, and monitored.

The impact of noise is currently taken into account by the Quebec Ministry of Transportation, which has recently adopted a formal noise policy. But previous and recent experiences with the implementation of noise mitigation measures on existing projects, and with the implementation of new projects of highway and urban boulevard show that any action, taking in consideration only the narrow perspective of noise mitigation, may be unsuccessful, creating local debates and political pressures. This is due to the fact that the proposed measures modify the quality of the environment of a local community which has wishes and apprehensions about the future of their neighborhood and which experiences their habitat with multiple senses. These previous experiences emphasize that the practice should move from the technical noise abatement to a more strategic and integrated (environmental and social) assessment of transport plans and policies during their development, performing these evaluations in an open and participatory process.

The aim of this chapter is to argue for a more comprehensive and strategic approach to transport policy. These new policies should be developed with a complete participation of the communities affected and concerned by these policies. This chapter is divided into three parts. First, we will summarize the traditional way to look at traffic noise control. Second, we will discuss the Quebec practice and experience in noise mitigation within Environmental Impact Assessment (EIA) processes. Third, we will argue that Strategic Environmental Assessment (SEA) provides the structure and means to look more broadly at these policies and their consequences.

THE TRADITIONAL APPROACH TO NOISE MITIGATION

The usual way to proceed in the analysis of noise related to a transportation project is consistent with a technical-rational model of decision-making. The decision to mitigate is taken according to a highly technocratic approach, which is specifically noise-oriented, even if esthetic, and other considerations are also taken into account (e.g., color of the noise barriers, landscape architecture).

In the context of mitigating existing infrastructures, this technocratic approach, founded on the recognition of the existence of a noise problem according to a certain norm and on the engineering solution to apply (noise barriers), is reactive. The managers wait for complaints and then, based on measurements, decide on the necessity to intervene. Any intervention requires the acceptance of responsibilities and frequently involves a sharing of the costs between organizations, for instance between the municipality and provincial ministries.

In the context of a new major project, EIA is generally required, and noise is always listed in terms of reference. The approach is corrective, and may involve public participation. The decision to mitigate is based on measurements and predictions of traffic, noise, and conceptual design (e.g., noise barriers, type of road surface). Limiting our discussion to noise impacts during the operation of the road, rather than at the construction stage, the technical — rational approach may be described as follows:

1) The baseline noise levels, the actual traffic, and its characteristics are measured for the affected area. The location of the receivers, the way they behave in their environment, and their sensitivity to noise are evaluated. The noise environment is qualified according to certain international standards or other regulations.

2) The expected noise environment is predicted through the modeling of noise dispersion according to different scenarios of traffic growth and characterization, landscape characteristics, meteorological data, and type of road surface. All these factors will have an effect on the amount and type of sound emitted, the way the sound will travel, and the eventual response of the receivers.

3) Actual and expected noise environments are compared. The changing situation is evaluated according to some noise significance criteria. A change of 3 dBA is considered as barely detectable; one over 15 dBA is considered to have an important impact.[2] The present level of noise is also considered to be of significance in the process.

4) A decision is taken according to the requirements of noise mitigation. If the expected noise exceeds the levels recommended, mitigation is proposed. Current mitigation strategies may include a reduction at the source (type of road surface, traffic management, road design, location of the infrastructures, etc.), a reduction by interference with the sound dispersion (noise barriers like fences and screens, buffer zone, etc.), or a reduction at the receiver level (site planning, architectural structure, house insulation, etc.).

5) The benefits of the selected mitigation strategy are evaluated, mainly in terms of cost/noise abatement ratio. Models are used to determine the characteristics of the noise barrier proposed (height, type of material,

location, etc.) and to evaluate the future situation with the mitigation measures implemented.

6) After the implementation, the noise environment is monitored mainly to validate the model applications. Rarely is the population's satisfaction evaluated.

The technical-rational approach to noise issue assumes that:

- The sound measurements are representative of the noise annoyance.
- All the people are equally affected by noise.
- Any action to reduce noise is beneficial to the society.
- In adopting some standard thresholds, equity may be reached at the regional level.
- For the noise barriers, it presumes that the hearing is more important than the other senses, including the sight.

THE QUEBEC POLICY AND PRACTICE

In Quebec, 135 km out of the 763 km of highways pass along residential areas where the traffic generates noise levels over 65 dBA.[3] The Quebec Ministry of Transportation (QMT), in its environmental policy, assumes its part of responsibility in resolving environmental problems related to transportation. In this policy, on the aspects related to noise or to the quality of life of residents living near highways, QMT undertakes:

- to integrate human and natural components within the Environment Assessment (EA) process;
- to apply restoration, mitigation, or compensation measures to improve or to exploit the environment;
- to apply measures to elaborate plans for it;
- to mitigate noise and other forms of pollution induced by the construction, use, and maintenance of transportation infrastructure;
- to evaluate and take into account the specificity of the area concerned by QMT interventions;
- to put in place mechanisms for public participation without the normal project development, and without taking into account whether the project may modify the quality of life;
- to enlarge the current practices of consensus-building and consultation.

At the project level, the QMT has adopted the technical-rational approach described above. Specifically related to noise pollution emerging from the Quebec road network, QMT[4-7] states that the solution is the erection of noise barriers. In order to improve the quality of life of the residents living near existing highways, QMT considers as a site requiring intervention on the request of a municipality, areas where the noise level is equal or over 65 dBA leq (24 h).[3] If it is so, noise barriers will be implemented if the city is willing to share (50–50) the costs of construction. These barriers will be designed to reduce the noise level by at least 7 dBA on the first row of houses bringing the noise level near 55 dBA, as long as the cost/effectiveness ratio of the barriers is acceptable.[3] For the QMT,[5] the solution to noise is mainly the construction of noise barriers, but other measures may be used such as: in the short term, the traffic management of heavy trucks, and in the long term, traffic redistribution, modifications to the structure of the roads, construction of new neighborhoods with higher tranquillity standards, and adoption of more severe architectural norms related to noise reduction. When new projects are considered, the criterion of 55 dBA or less for sensitive areas is applied; but, where the sound level is already exceeding 55 dBA, the actual level will be referred to as the standard.[7]

THE PUBLIC EXPERIENCE, NOISE, AND EIA

How do people feel and react to the technical-rational approach? To evaluate feelings and reactions, 32 public hearings and mediation reports produced by the Bureau d'audiences publiques sur l'environnement (BAPE) between 1980 and 1996, have been examined by the authors of this presentation through a content analysis.

In general, noise was considered the main issue. The members of the commissions and the citizens referred to noise as a dimension of the overall quality of life, or as an agent of degradation of a healthy environment. The multidimensional structure of quality of life has been discussed only once specifically in a public hearing.[8] The commission reminds us that EIA always implies judgments, and that impacts of different natures will affect a person synergistically, not separately. The commission states that it is the cumulative impacts experienced by one person, even if these impacts are minor, that may affect the quality of life. The intensity of this person's perception of how they will be impacted depends on the person's own evaluation of the significance of the different impacts, and also on the person's feeling toward the project.

The report analysis has permitted us to identify the chain of causality generally in the minds of people. People react drastically to the risk of change in their environment, an environment they have chosen for different reasons, including the tranquillity; people dislike less the sounds that they have voluntarily chosen than those which are imposed on their environment.[11] When a new project is announced, affected people generally make the following interpretation:

With this project, the traffic will increase. This increase will generate higher levels of noise, vibration, and dust. These levels of noise, vibration, and dust will deteriorate our quality of life (housing and neighborhood), and reduce our property value. In the case of an apartment building, these impacts will increase the level of vacancy, increase the turnover of locators, generate a diminution of the rents, and reduce investments in the neighborhood.

When a risk of change is perceived, the analysis shows that people generally question the justification of the projects, the validity of the measured and predicted traffic levels and composition, and the noise measurements and estimates. The value of the leq 24 h, the period of time between the baseline study and the public hearing, or the measurement strategy (duration, number of measurements, period of the days or of the year, etc.) are usually contested. Many affected residents seem willing to consider an expropriation rather than undergo an expected deterioration of their quality of life,[12,13] while people subjected to expropriation generally fight against it.

Many different approaches to fight against noise problems have been proposed : traffic management, better road maintenance conditions, modification of design criteria, relocation of houses or people, noise barriers or screens. Very few reports have mentioned modifications of buildings.[8,16] Even when all the precautions are taken, if the noise level is still perturbing, people keep their windows closed, or relocate their activities within their apartment (sleeping, reading, etc.); the main portrayal of this situation is the image of imprisonment.[17]

Noise barriers were planned by QMT for the first time within the life cycle of a project in 1987.[18] When noise barriers are recommended, the discussion is oriented toward their efficiency, visual appearance, impacts on the landscape, and conse- quences for the residential environment (e.g., dust, microclimatology). Relative importance of noise impacts and visual impacts have been debated. Harmonization between the senses of hearing and sight is reached by slight modifications of the highway location, or by landscape architecture practices. Noise barriers sometimes appeared to deteriorate quality of life through impacts on property value, and on vacancy rates in apartment buildings.[19] The main portrayal of living near a noise barrier is the image of confinement; people feel their neighborhood confined within a certain area. It seems that the perceived noise reduction is less than the effective measured reduction.[20]

On different occasions, some ethical questions of justice and equity have been discussed. These matters appeared when the project implied a redistribution of the traffic patterns, and when noise barriers are proposed in medium and high-density areas. In the case of traffic redistribution, the problem may be summarized by a few questions such as: why bring the noise problem in our backyard? Could we reallocate the traffic in such a way that the impacts will be shared?[21,22] In the case of noise barriers, the questions are: because the noise barriers are efficient only at the ground level, is it equitable to redistribute and channel the sound in a way that the noise will increase from the second story to the top of a building? Because there does not appear to exist clear solutions for multi-storied buildings, is it possible to think of some compensation, even if the causal relation between the rate of vacancy, the reduction of the monthly rent, and the increase of noise is not easily established?[13,19,22]

Public participation is well received by the citizens. Because of the importance of impacts of transportation projects on the daily life of affected people, the propo- nent of a project should meet the affected people early in the EIA process to discuss with them openly and with transparency the impacts, trying to jointly identify a strategy to improve their general well-being.[13,14] When uncertainties persist, it was sometimes proposed to create round-tables or committees to conduct the follow-up and to propose modifications to the mitigation strategies, if necessary with the commitment of the lead agency.[22]

Finally, on many occasions, different commissions have mentioned that priority must be given to prevention (e.g., a good land-use and zoning approach) through plan and regulation.[20] *A posteriori* mitigation, if necessary, were preferred by some commissions to any kind of compensation.

FROM PROJECT EIA TO STRATEGIC ENVIRONMENTAL ASSESSMENT

LESSONS FROM PROJECT ENVIRONMENTAL IMPACT ASSESSMENTS

Dealing with noise impact within project EIAs, the previous discussion suggests that:

- Noise is a major transportation-related impact;
- Noise is only one dimension within a perspective of a multidimensional quality of life approach;
- People are unequally affected by the noise according to their location, and to their different physiological, psychological, cultural, social, and economical characteristics;
- Sound level criteria are debatable and must reflect sound dynamics and periodicity;
- High sound levels and changes in sound exposure may cause health problems and generate different ways, mainly behavioral, to cope with the problem;
- The adoption of a standard strategy of mitigation, like the noise barriers, applied independently of the location specificity, may generate inequities;
- Public involvement in the on-going process of transportation planning (from the early conception of the regional plan to monitoring of a project) is required.

The problem of noise is more than an issue for experts and scientists. Even if measurements and predictions are useful and necessary, the question needs to be addressed not only as an engineering problem to be resolved, but as a problem of local development. The publics understanding of the problem is based on values rather than on science. Noise being only one of the transportation-related problems among other cumulative impacts, the global issue of quality of life needs to be addressed. Why should we plan noise barriers when, after implementation at a cost varying between CDN \$500.000/km and CDN \$1.200.000/km, people are frequently unsatisfied with the perceived noise reduction and with the general appearance of infrastructures, and are complaining about their visual obstruction?[23]

THE NEED FOR STRATEGIC IMPACT ASSESSMENT

The technical-rational approach with a low public involvement and the application of rigid noise criteria for intervention fails to satisfy the public, even if it simplifies decision-making. People generally feel that their values and opinions have to be taken into account throughout the planning stage and within the decision-making process. When the participation is concentrated at the public hearing level, people are forced to contest scientific facts: measurements, predictions, efficiency, etc.

This analysis of transport project EIA emphasizes the need for a more global assessment of transport plans and policies. It appears clearly that:

1) Transport means should be tied to consensus-based development.
2) Alternative approaches to transport should be examined from a regional perspective and public participation with all the communities may yield an alternative that is not a highway or an urban boulevard.
3) If highways or urban boulevards are the preferred alternative, then compatible land uses should be the focus of future development.
4) Planners should realize that the means of transport forecloses development alternatives.
5) If noise interventions are required, the project assessment should move from a purely technical-rational approach, to a more strategic environmental and social assessment of the quality of life, in such a way that the overall quality of the neighborhood development will be improved.

In this period of economic austerity, more efficiency is required in dealing with transportation-related problems. SEA is a tool which may be useful for achieving efficiency. In this context, SEA with public involvement, needs to be applied to:

- the regional development plans and strategy
- the regional transportation plans
- the programs to improve the general quality of life in relation to transportation

Moreover, each specific project should be considered within its own specificity.

SEA Implementation, Victim of System Inertia

Within its environmental policy, the QMT seems open to the realization of participatory SEA: the will to improve the quality of the environment, the commitment for a better public involvement, the search for consensus-building, etc. The inertia in the system is related to:

- the lack of will to share power in decision-making
- the inter-agency and intergovernmental distribution of jurisdiction (e.g., urban planning, regional planning, environmental protection, etc.), and the risk perceived by these organizations of loss of authority
- the dominance of the usual technical-rational approach

We should also emphasize that the Quebec law on environmental quality does not specify the obligation to submit plans, programs, or policies (PPP) to any environmental evaluation. It is still limited to major projects and to a few programs of activities like dredging and pest control. But the minister of Environment may require some public hearings on questions of public major interests. During the passed years, he required public hearings for hazardous waste management, residual material, and forest management; these can be considered good steps toward participatory SEA.

Promoting SEA

The EIAs performed for transportation infrastructures in Quebec during the last 30 years have contributed significantly to modify some governmental practices. The public interventions have stimulated a modification of noise measurement procedures; the experts are now looking at sound peaks and frequency distributions by day and night, and reporting to the concerned public. The QMT has produced a book explaining noise concepts, measurements, and management for those concerned and interested.[9] Projects managers are now involving people earlier within the project evaluation process; the use of noise simulation and photographic compositions allows people to perceive the modifications that the project will generate on their local environment and to comment and react to the project or to the proposed mitigation measures. Their environmental communication strategies have been greatly improved.

But, project EIA still do not address some major societal concerns. Even if people can stop a project and debate on its justification or influence the noise mitigation strategy, they cannot really influence the development or application of the regional transportation plans or strategies which have been approved previously by the government. They cannot participate significantly in the debate on the urban sprawl and development. The questions of justice and equity related to the alteration of their quality of life, including the quality of their environment, cannot be addressed significantly. Finally, the process is mitigation-centered rather than prevention-centered.

We think that participatory SEA will contribute to:

- the reduction of expenses for mitigation measures
- a more cost-efficient allocation of the state's investment
- the improvement of the proactive image of the government
- the state's commitment to a democratic, open, and transparent development process
- a synergy between concerned agencies, governments, other organizations, and citizens towars the improvement of our environment
- project EIA being focused on local issues rather than societal debates, thus reducing confrontations between actors.

ACKNOWLEDGMENT

The authors would like to thank Professor Peter Boothroyd, Centre for Human Settlements, University of British Columbia, for useful and constructive comments on this chapter.

REFERENCES

1. André, P. and Gagné J.P., *Atténuation du bruit routier en milieu résidentiel dense, Première étape: Revue et analyse de la littérature (rapport final)*, Rapport de recherche produit pour le ministère des Transports du Québec, Montréal, 1997.

2. Therivel, R., Noise, in *Methods of Environmental Impact Assessment*, Morris, P. and Therivel, R., Eds., UBC Press, Vancouver, 1995, 4.

3. QMT, *Le bruit causé par la circulation routière, Orientation ministérielle préliminaire*, Gouvernement du Québec, ministère des Transports, Québec, 1995.

4. Québec, *La politique sur l'environnement du ministère des Transports du Québec*, Gouvernement du Québec, ministère des Transports, Québec, 1994.

5. Québec, *Éléments de problématique et fondements de la politique sur l'environnement du ministère des Transports du Québec*, Gouvernement du Québec, ministère des Transports, Québec, 1994.

6. Québec, *Les orientations du gouvernement en matière d'aménagement, Pour un aménagement concerté du territoire*, Gouvernement du Québec, Québec, 1994.

7. Québec, *Normes – ouvrages routiers, Gouvernement du Québec, ministère des Transports*, Québec, 1994.

8. Bureau d'audiences publiques sur l'environnement, Projet de réaménagement de la route 116, tronçon Princeville/Plessisville, Gouvernement du Québec, Rapport d'enquête et d'audience publique no. 21, Québec, 1986.

9. Québec, *Combattre le bruit de la circulation routière. Techniques d'aménagement et interventions municipales*, Les Publications du Québec, Québec, 1987.

10. Bureau d'audiences publiques sur l'environnement, *Autoroute 55: doublement de la chaussée entre Bromptonville et l'intersection avec le chemin de la Rivière*, Gouvernement du Québec, Rapport d'enquête et de médiation no. 71, Québec, 1993.

11. Bureau d'audiences publiques sur l'environnement, *Prolongement de l'autoroute 73 vers Stoneham*. Gouvernement du Québec, Rapport d'enquête et d'audience publique no. 28, Québec, 1988.

12. Bureau d'audiences publiques sur l'environnement, *Réaménagement de la route 112-116 entre les échangeurs Charles-Lemoyne et Saint-Hubert, incluant l'échangeur Edna-Maricourt*, Gouvernement du Québec, Rapport de médiation no. 62, Québec, 1993.

13. Bureau d'audiences publiques sur l'environnement, *Projet de construction de l'échangeur Brière sur l'autoroute 15 et d'une voie de desserte (Saint-Jérôme - Bellefeuille)*, Gouvernement du Québec, Rapport d'enquête et d'audience publique no. 100, Québec, 1995.

14. Bureau d'audiences publiques sur l'environnement, *Construction du tronçon de l'autoroute 50 entre Lachute et Mirabel*, Gouvernement du Québec, Rapport d'enquête et d'audience publique no. 35, Québec, 1990.

15. Bureau d'audiences publiques sur l'environnement, *Projet de réaménagement du chemin de la Montagne dans la Ville de Hull*, Gouvernement du Québec, Rapport d'enquête et d'audience publique no. 23, Québec, 1987.

16. Bureau d'audiences publiques sur l'environnement, *Projet de réaménagement de la route 307 entre les ponts Alonzo-Wright et des Draveurs à Gatineau*, Gouvernement du Québec, Rapport d'enquête et d'audience publique no. 51, Québec, 1992.

17. Bureau d'audiences publiques sur l'environnement, *Projet d'urbanisation de la route 173 à Saint-Georges de Beauce*, Gouvernement du Québec, Rapport d'enquête et d'audience publique no. 26, Québec, 1988.

18. Bureau d'audiences publiques sur l'environnement, *Projet de prolongement de l'autoroute 19, de l'autoroute 440 jusqu'au boulevard Dagenais à Ville de Laval*, Gouvernement du Québec, Rapport d'enquête no. 23b, Québec,1987.

19. Bureau d'audiences publiques sur l'environnement, *Réaménagement de la route 337 de l'autoroute 640 au chemin Martin Newton*, Gouvernement du Québec, Rapport d'enquête et de médiation no. 91, Québec, 1995.

20. Bureau d'audiences publiques sur l'environnement, *Implantation d'une voie d'accès au secteur nord de Trois-Rivières*, Gouvernement du Québec, Rapport d'enquête et d'audience publique no. 57, Québec, 1993.

21. Bureau d'audiences publiques sur l'environnement, *Projet de réaménagement de l'échangeur de l'autoroute 15 au kilomètre 31, à Mirabel (Saint-Janvier)*, Gouvernement du Québec, Rapport d'enquête no. 27, Québec, 1988.

22. Bureau d'audiences publiques sur l'environnement, *Projet d'établissement d'un dépôt de matériaux secs à Saint-Pie*, Gouvernement du Québec, Rapport d'enquête et d'audience publique no. 93, Québec, 1995.

23. Ouimet, G., *Acceptation sociale d'écrans acoustiques en bordure d'autoroutes, le cas de Ville de Laval (Québec)*, M.Sc. Thesis, département de Géographie, Université de Montréal, Montréal, 1994.

Section VI

Forum

19 The Future of SEA

Riki Therivel and Maria Rosário Partidário

CONTENTS

INTRODUCTION

The previous chapters have identified a wide range of perspectives on Strategic Environmental Assessment (SEA), both theoretical and practical. They give examples of SEAs carried out in many different countries (including some very innovative work from "emerging" nations), by different types of organizations, under different circumstances. Happily, they show a high degree of consensus on many of the important issues underlying SEA, as well as displaying some heartening and inspiring examples of good practice. On the other hand, some of the assumptions that underlie these chapters are unclear or worth challenging. It is also worthwhile checking whether, under current trends, future SEA systems will be as "good" as we would wish them to be.

This chapter begins by summarizing two of the main themes emerging from the previous chapters: the need for SEA, and examples of good practice SEA. It then considers three themes where the messages seem to be more confused or still emerging: planning-SEA links, sustainability-SEA links, and the effectiveness of SEA. It concludes with brief recommendations for future SEA practice.

SEA IS NEEDED

Many of the previous chapters highlighted the *need for SEA*, often in considerable detail (e.g., Clark, Shuttleworth, and Howell, Bass and Herson, Machac et al.). SEAs can:

1-56670-360-3/00/$0.00+$.50
© 2000 by CRC Press LLC

- improve the proactive and green image of the government or corporate body, increasing the acceptability of their proposals
- increase the professional expertise of the staff carrying them out
- internalize the environmental values of all the people involved in the planning process
- avoid unnecessary environmental, social, and economic costs (for example, unnecessary or ineffective mitigation or compensation costs)
- avoid unnecessary delays because they decrease the chance of litigation
- improve the planning process by identifying weaknesses and deficiencies

Shuttleworth, in particular, stressed the need to ensure that decision-makers and those carrying out SEAs understand these benefits: "... SEA is not something that is just what we are already doing."

The chapters also give some poignant examples of limitations of project Environmental Impact Assessment (EIA). These include the cumulative impacts of the South African Saldanha Steel Project (Wiseman), the "island" of residents left between a Czech road and railway line (Machac et al.), and the inability of EIA to deal effectively with noise impacts (André and Gagné).

Effective "tiering" of SEA and EIA is widely seen as a good framework for EIA. As Clark stressed, "SEA is a new way of looking at EIA, a new way of looking at when it is prepared, a new way of looking at the scope of its analysis and a new way of looking at the level of detail required for an EIA".

Within strategic decision-making there are different levels, and the integration of environmental concerns should take place (in a tiered approach) at all these levels. Although SEA may have the greatest benefits at the policy level, it may be wise to also initiate SEA at more pragmatic levels of programs and plans, especially in countries where policy-making is not yet fully endorsed. Examples of successful tiering include water provision in the Netherlands (Verheem) and Spain (Brooke), regional/local planning in Tonga (Morgan and Onorio), and park management in Canada (Therrien-Richards).

GOOD PRACTICE EXAMPLES OF SEA EXIST

The chapters do not show a clear agreement on what is "good SEA practice," nor how such practice can be achieved. To an extent, what is effective SEA will depend on the specific context in which it is applied. Thus, *SEA should be seen as a range or 'family' of tools* rather than as one specific approach (Brown and Wiseman). However, more universal agreement is needed on the basic principles, standards, and terminology in order to be able to sell the concept of SEA to its potential users: policy and planning decision-makers.*

---

* Confusion about what SEA is will not help in convincing skeptical decision-makers. An important group of decision-makers are politicians. It is important to realize that politicians will only accept SEA if it enables them to demonstrate to their voters the soundness and democratic character of their decisions. This will increase their credibility.

Many of the previous chapters highlight examples of inspiring, innovative practice that have proven effective in at least their context. Examples include, broadly in the chronological sequence of SEA:

- Formal *screening* of strategic actions that require SEA in Canada (Therrien-Richards) and California (Bass and Herson).
- Use of *sustainability indicators* in regional drinking water management in the Netherlands (Verheem), the national energy policy of the Czech Republic (Machac et al.), for Canadian international policies and programs (Shuttleworth and Howell), and by various international organizations (Partidario and Moura). Note that all four are sustainability, not just environmental, indicators. The first two also use weightings.
- Comparison of *alternatives*: for instance different approaches to meeting water demand in California's East Bay District (Bass and Herson) and in the Netherlands (Verheem); to a national energy policy in the U.S. (Clark) or in the Czech Republic, including one developed through public participation (Machac et al.); and to delivering strategic objectives under the Canadian Farm Income Protection Act (Hazell and Benevides). The "no action" alternative was explicitly considered in the Californian and Canadian studies, while demand-side management — arguably a key alternative in any strategic action related to water provision — was considered in both water-related SEA examples.
- *Methods of analysis and evaluation of impacts*, including the compilation of "vulnerability maps" in the Netherlands using Geographical Information Systems (GIS) (Verheem); the division of Canadian parks into zones that reflect their vulnerability and test the strategic actions against these zones (Therrien-Richards); use of GIS to create and analyze information systems for the KwaZulu Regional Economic Forum (Wiseman); use of maps and map overlays to identify cumulative impacts in Canada (Therrien-Richards); and algorithms for ranking/selecting environmental restoration projects in Louisiana (Wilson and Hamilton).
- *Mitigation measures*. Some of these related to lower-tier SEA/EIA, for instance, the preparation of "second-tier" SEAs on the selected alternative for water supply in California (Bass and Herson), and specific project-level EIAs for Tonga. Others focus on management of subsequent projects. In South Africa and Tonga, the case study SEAs resulted in detailed guidelines for planning, EIA, and environmental management at the implementation stage (Wiseman, Morgan, and Onorio). In Canada, SEAs suggested the use of best practice during project implementation, for instance the use of best practice guidance in building trails (Therrien-Richards).
- Some case studies highlighted examples of innovative ways for *involving the public and other stakeholder groups* in the SEA. In the Czech Republic, all important information, results, and meeting dates were published on a web page, and teleconferencing with non-governmental organizations took place (Machac et al.). For the Canadian Net Income Stabilization

Account, questionnaires were distributed to 10,000 farmers to identify relevant environmental issues (Hazell and Benevides).

- Finally, *monitoring* was generally not yet carried out, but relevant indicators had been established in Finland for European Union Structural Fund allocation (Törttö), and the SEAs carried out under Canada's Farm Income Protection Act are themselves partial monitoring studies, being carried out after any agreement is passed (Hazell and Benevides).

Overall, then, within the framework of SEA thought, the case studies and the conference on which they were based show broad agreement on the need for SEA and suggest some techniques for how "good" SEA could be carried out. Now three more contentious issues are raised.

WHAT ARE LINKS BETWEEN PLANNING AND SEA?

Healey et al.[1] suggest that approaches to planning (in Britain at least) are moving from two main models toward two other models, as shown in Figure 19.1.

Rationalist decisionmaking: planners are analyzers and regulators who convert their clients' (politicians', the public's) values into goals/criteria, and collect and analyze "scientific" information based on these criteria.	↘	↗	Collaborative consensus-making: planners are facilitators and mediators who aim to involve a wide range of stakeholders in decision-making, using, e.g, development of institutional capacity, dialogue, "visioning," "planning for real," parish mapping, etc.
		now	
"Negotiative" planning: a main trend of the 1980s and early 1990s, focuses on *enabling* activities (particularly private sector activities) as opposed to *providing* state-funded services and infrastructure. It places the planner in the role of negotiator and adjustor, and takes its most visible form as bargaining over development projects' "planning gain."	↗	↘	"Neoliberal technical regulation": planners are rule-setters and administrators who set a context for private sector enterprise through, e.g., standards, zoning regimes, rules for developer contributions, and fiscal measures.

FIGURE 19.1 Approaches to planning (based on[1,2]).

The previous chapters clearly exemplify three of these approaches and their links to SEA. Many SEAs as currently practiced fall squarely within the tradition of *rationalist decision-making*, with technocratic decision-making and limited opportunities for public participation. The most distinctive of these is the Wetlands Value Assessment model used to select wetlands restoration projects in Louisiana. This was based on the views of experts (only) and are based on only ecological criteria. The Canadian SEA processes described by Hazell and Benevides generally involved some form of public "consultation," but were essentially carried out by planners or environmental consultants. In Quebec, noise impacts are dealt with by the Ministry of Transportation, again using technical-rational approaches with public views emerging at public hearings and mediations (André and Gagné). Wiseman suggests that planning approaches in South Africa are moving from zoning (purely technocratic), to indicators of preferred trends and development types, to approaches that incorporate public participation and empowerment.

Evolving best practice approaches to SEA encourage broad public involvement in the framing of goals/criteria, choice of alternatives and "ownership" of development projects, reflecting the emerging *collaborative consensus-making* approach to land use planning. Morgan and Onorio suggest that planning should take place "with people" not "for people." Törttö promotes the concept of "planning as learning." Brown notes that "participatory discussion... on problems and possible solutions fits more comfortably with the way conflicts are resolved... than do the more analytical and aloof processes provided by EIA-type approaches." They suggest that participation by the public and other relevant (generally more technically expert) groups can:

- bring local knowledge that can improve the design and evaluation of options
- better reflect the plurality of interests and environmental values involved in the development of the strategic actions
- identify mitigation measures to better address peoples' real concerns
- help to ensure that strategic actions are effectively implemented
- educate all those involved

On the other hand, some case studies display SEA in an uncomfortable role, where best-practice collaborative SEA approaches conflict with an overriding *neoliberal technical regulation* approach to planning. Morgan and Onorio's review of the Neiafu Master Plan, for instance, notes that the focus of the plan, which was prepared by an international consultancy, "was to be on improving the living standards of local people... based on a range of development proposals," which were mostly for infrastructure and tourism. Yet, they note, "it is not clear... to what extent the local community... participated in the development programme... and there is no indication that participation ought to be a consideration in the continuing programme." Partidário and Moura's example of an indicator, trip length for daily commuting, is clearly aimed at measuring individuals' response to a policy setting, again an example of neoliberal technical regulation approach. Törttö's study involves applying SEA to regional development programs: there, "environmental impacts... are hard to assess because information concerning these projects is considered to be commercially sensitive... Project selection criteria... usually are related to economic activity." Brooke suggests that, in Spain,

> hydrological planning continues to be driven by the need to meet demand, especially irrigation... [the plan] include[s] the requirement for an 'environmental' minimum flow. However this is treated as an operational condition rather than a water use constraint, and so is sub-ordinate to demand use. The plan has clear recognition of the need for economic appraisal of 'project' options.

In sum, SEA practice is moving from a purely rationalist decision-making approach, as is the case with EIA, to one which reflects "collaborative consensus-making" approaches to land-use planning. However, it sits uncomfortably with neoliberal technical regulation approaches to planning. The question here may be: is this the fact that is preventing SEA from being more easily accepted and adopted? Is it because it is being seen as another highly regulated procedure which creates more complexity in the decision-making process given the strategic level at which it applies? In such a case a major priority with SEA may well be the search for ways to conceptualize and implement SEAs that fit better with existing decision-making structures at policy, planning, and program (PPP) levels. Rather than a regulation, SEA should probably be no more than best practice, translated into adequate guidelines that can assist policy-making, planning processes, and program development in a cost-effective manner.

WHAT ARE LINKS BETWEEN SUSTAINABILITY AND SEA?

Approaches to sustainability can be placed on two different, interrelated scales. One scale describes "greenness:" whether environmental constraints are seen as overriding (dark green) or as being offset by economic or social gains (light green/"ecological modernization"). Visually, the dark green approach could be represented as:

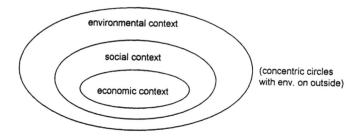

The light green approach could be represented as:

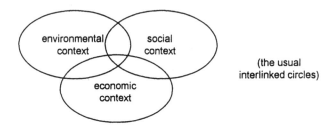

(the usual
interlinked circles)

The other scale denotes the links between economic issues and quality of life (QoL): whether QoL is seen as being equivalent to economic growth, related to economic growth, but also incorporating a social dimension, or independent of economic growth. Main approaches to sustainability/environmental management according to this model are shown at Figure 19.2.

Because SEA is so inextricably linked to the decision-making process, depending on the context, SEA can fit into most of the cells of Figure 19.2. Most of the SEAs described in previous chapters are tools for "ecological modernization" that reflect the Brundtland Commission's approach to sustainable development, which suggests that environmental protection and economic development can only progress hand in hand (i.e., the top left corner of Figure 19.2). Under this model, SEA is used to fine-tune strategic actions and mitigate their most negative environmental consequences. For instance, the "sustainability" targets discussed by Partidário and Moura show a combination of economically "feasible" environmental targets (coastal zones should not be depleted by more than 10% of the total area) and social targets regarding noise and risk. Törttö refers to indicators on "increase in tourism" and "creation of small and medium-sized enterprises." In at least the Spanish and the Schiermonni-koog examples (Brooke, Verheem), water use was expected to increase by more than 50% over the life of the plan, with severe environmental repercussions, but with little emphasis on demand reduction. Cynically viewed, this type of SEA allows decision-makers to enter the heaven of environmental righteousness after having survived the purgatory of writing an SEA report. On the scale from light to dark green, SEA as it is normally and currently practiced, is so light-green as to be virtually invisible.[4]

However, SEA has the potential to be a much stronger tool for "dark green" sustainability by incorporating "true sustainability" targets or carrying capacities into strategic actions, and arguably by challenging the links between quality of life and economic growth (i.e., by moving from the top left toward the bottom right corner of Figure 19.2). Unfortunately, not one of the case studies uses such targets. Several of the case study SEAs — notably those of the Netherlands and California (Verheem, Bass) — did consider demand reduction as an early alternative, which was later dropped, and Verheem discusses the need for clearly-phrased goals. However, Brooke points to the problem of environmental objectives being incorporated in early stages of the development of strategies, but being dropped or watered down by the time the ultimate strategic action is prepared; and Brooke and Verheem both

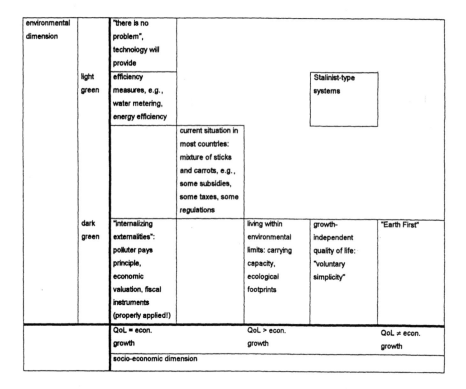

environmental dimension		"there is no problem", technology will provide				
	light green	efficiency measures, e.g., water metering, energy efficiency			Stalinist-type systems	
			current situation in most countries: mixture of sticks and carrots, e.g., some subsidies, some taxes, some regulations			
	dark green	"internalizing externalities": polluter pays principle, economic valuation, fiscal instruments (properly applied!)		living within environmental limits: carrying capacity, ecological footprints	growth-independent quality of life: "voluntary simplicity"	"Earth First"
		QoL = econ. growth		QoL > econ. growth		QoL ≠ econ. growth
		socio-economic dimension				

FIGURE 19.2   Approaches to sustainability. (Based on Ref. 3. With permission.)

refer to cases where environmental goals are stated, but are very broad, while economic goals are much more specific and workable.

A shift to a "darker green" approach to sustainability that integrates social concerns more comprehensively would require a substantial rethinking of the approach toward SEA. In particular, it would need to actively take on board Wiseman's dictum that SEA should focus on the effect of the environment on development, not development on environment. We perceive this as being one of the main challenges for SEA in the new millenium.

WHAT IS EFFECTIVE SEA?

In his chapter, Thissen establishes a clear and useful set of criteria for evaluating whether any policy analytic activity, including SEA, is effective. Arguably the key criterion is whether the SEA changes/improves the strategic action, i.e., its "use" and "effects." Although most of the case studies described here are unclear about the effects of the SEA on the strategic actions, several case studies give examples of SEAs that had an impact on decision-making, for example:

- In the Netherlands, "the SEA did influence the decision-making process" for the National Plan on Drinking and Industrial Water, and for drinking water production on the island of Schiermonnikoeg the two most environmentally friendly options were chosen (Verheem).

- An SEA-type process has been used to distribute almost $250 million to wetlands restoration projects in coastal Louisiana (Wilson and Hamilton).

However, only in the example of Louisiana (Wilson and Hamilton) was enhancement specifically referred to, and arguably, this was in response to decades of environmental harm that also has considerable economic ramifications.

Case studies, and discussions with SEA practitioners and researchers suggest that effective SEA is often linked to:

- starting the SEA early (Thissen's formal context of activity, availability of time)
- involving a range of disciplines and the public (willingness/availability of actors, extent of cooperation)
- approaching the SEA in an open-minded manner, and seeing it as an educational process
- being transparent about the appraisal process and results
- focusing on enhancement rather than just mitigation

WHAT IS THE FUTURE OF SEA?

The previous sections suggest that SEA has a bright future. SEA is increasingly seen as a way to counter limitations of EIA and help promote sustainability. Good practice techniques are proliferating, particularly more participative techniques, the use of GIS in SEA, and the comparison of alternatives.

Appropriate forms of SEA are likely to be strongly influenced by the shape of future planning. Where planning systems evolve toward a "collaborative consensus-making" model, SEA will be increasingly seen as a tool for public participation and improved transparency of decision-making. This could be done by incorporating aspects of "visioning," involving local people in the development of alternative strategic actions, and linking SEA with Local Agenda 21 approaches. Where, instead, planning systems evolve to a more "neoliberal technical regulatory" approach, many aspects of decision-making are likely to shift from a public to a private arena, thus influencing the transparency of decision-making, the balance of economic, social, and environmental concerns, and the role of SEA. SEA would probably be carried out in fewer circumstances and/or would take the role of an anodyne "green check" carried out by private bodies for marketing purposes. Where industry's need for commercial confidentiality conflicts with transparency of decision-making, new procedures for public participation and dissemination of SEA results will be required. New techniques and procedures for SEA of high-tier strategic actions such as privatization, international trade agreements, and new technologies will also be needed.

SEA, in the future, will also be affected by different scales of planning. The trend toward globalization means that a given country's strategic actions are increasingly influenced by those of other countries. New protocols will be needed for SEA in a transboundary context, as well as for translating multinational agreements into targets for individual strategic actions (Schramm). Trends toward regionalization will result in the formation of new regional-level agencies and new approaches to regional-level SEA. Relevant issues include the roles of various regional agencies, appropriate regional-level targets and indicators, relevant data and models, and sector-specific appraisal techniques.[4]

In terms of evolving approaches to sustainability, much of the potential effectiveness of SEA depends on the establishment of environmental/sustainability targets.[5] Further research is needed to identify targets that are relevant to SEA, and especially to determine how they can be disaggregated from a national to a regional/local level. SEA is also most likely to promote sustainability if it is fully integrated into strategic decision-making. However, further research may be needed to test whether the advantages of integration outweigh the possible disadvantages of diluting the "auditing" role of SEA.[4]

In the final analysis, SEA should improve environmental conditions. Carrying out an SEA means using electricity, paper, petrol, and food, to run the brains of the people carrying it out. We need to show something in return. At the very least, the result of an SEA should make up for those "inputs." Ideally, SEA would help us to move toward a fairer, cleaner, more diverse world where wealthy countries do not "lean" on poorer ones, and where decisions are so environmentally sound that there is no more need for SEA.

REFERENCES

1. Healey, P., Khakee, A., Motte, A., and Needham, B., Eds., *Making Strategic Spatial Plans: Innovation in Europe*, UCL Press, London, 1997.
2. Bruton, M. and D. Nicholson, *Local Planning in Practice*, Hutchinson, London, 1993.
3. CAG Consultants, *Background Paper 5: Sustainability*, report for the Royal Commission on Environmental Pollution Study on Energy and the Environment, London, 1998.
4. Therivel, R., *Strategic Environmental Assessment: From Theory to Practice*, Ph.D. thesis, Oxford Brookes University, Oxford, 1999.
5. Partidário, M.R., "Strategic environmental assessment: key issues emerging from recent practice," *Environ. Impact Assess. Rev.*, 16, 31, 1996.

Index

Printed and bound by CPI Group (UK) Ltd, Croydon, CR0 4YY

23/10/2024

01777670-0013